Contemporary's

Number Power

a real world approach to math

Robert Mitchell and Donald Prickel

McGraw Hill **Wright Group**

Wright Group

Printed in the United States of America.

Send all inquiries to:
Wright Group/McGraw-Hill
130 E. Randolph, Suite 400
Chicago, IL 60681

ISBN 0-8092-2381-3

12 13 14 15 16 CUS 09

The **McGraw·Hill** Companies

Acknowledgments

(graphs) "National Football League Revenue Sources" and "National Hockey League Revenue Sources." From *The World Book Year Book.* © 1996 World Book, Inc. By permission of the publisher. http//www.worldbook.com

(graph) "World Population to Soar." From *The World Book Year Book.* © 1995 World Book, Inc. By permission of the publisher. http//www.worldbook.com

(map) "Where Asian Americans Live." From *The World Book Encyclopedia.* © 1999 World Book, Inc. By permission of the publisher. http//www.worldbook.com

(chart) "Recommended Daily Dietary Allowances." From *Recommended Dietary Allowances,* 1995, Ninth Edition, National Academy of Sciences.

(chart) "Daily Food Guide." From National Heart, Lung and Blood Institute, NIH publication #94-2920. As appeared in *Step by Step: Eating to Lower Your High Blood Cholesterol,* American Heart Association, 1995.

(graph) "Guns in the Home." From National Opinion Research Center, University of Chicago, 1998, as appeared in *Wall Street Journal,* March 12, 1999, page A6.

(chart) "Traffic: It's Going to Get Worse." From Lane Council of Governments, as appeared in *The Register Guard,* March 7, 1999, page 16a.

Table of Contents

Schedules and Charts

Maps

USING NUMBER POWER   163

To the Student

Welcome to *Graphs, Charts, Schedules, and Maps*:

Number Power 5 introduces you to the skills you need to read a wide variety of graphs, charts, schedules, and maps. An understanding of these topics is necessary in many occupations and in your everyday life. Also, reading graphs, charts, schedules, and maps is a standard section on educational and vocational tests, including GED, college entrance, civil service, and military tests.

The first section of this book, Building Number Power, provides step-by-step instruction in reading and interpreting these kinds of materials. You will learn to find information from four kinds of graphs: pictographs, circle graphs, line graphs, and bar graphs. You will practice reading a variety of schedules and charts that you might face on a daily basis. And you will learn how to interpret information from three types of maps: geographical maps, directional maps, and informational maps.

Each chapter begins with a skills inventory to help identify what you need to learn. After completing the lessons and exercises, you will be challenged in applying those skills to everyday problems. Each chapter ends with a skills review to check your progress on what you have just learned.

In addition, several skills that you will find helpful in your study of graphs, charts, schedules, and maps are calculator use, mental math, and estimation. These skills are reviewed on pages 191, 192, and 194. Also, for quick reference, a summary of measurement conversions and a glossary of math terms can be found at the back of this book. The following icons will alert you to problems where using a calculator, estimation, or mental math will be especially helpful.

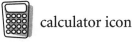 calculator icon

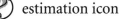 estimation icon

 mental math icon

To get the most out of your work, do each problem carefully. Check each answer to make sure you are working accurately. An answer key is provided at the back of the book. Inside the back cover is a chart to help you keep track of your score in each exercise.

Pretest

This test will tell you which sections of *Number Power 5* you need to concentrate on. Do every problem that you can. Correct answers are listed by page number at the back of the book. After you check your answers, the chart at the end of the test will guide you to the pages of the book where you need work.

GRAPH A

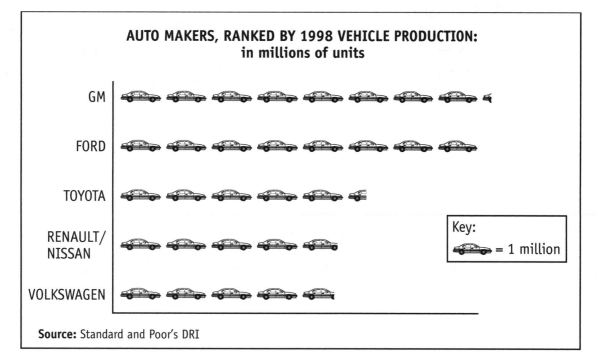

AUTO MAKERS, RANKED BY 1998 VEHICLE PRODUCTION:
in millions of units

GM

FORD

TOYOTA

RENAULT/
NISSAN

VOLKSWAGEN

Key:
= 1 million

Source: Standard and Poor's DRI

1. Graph A shows vehicle production in _____, with the symbol representing _____ cars.

2. In 1998 Toyota produced almost $5\frac{1}{2}$ million cars.   True   False

3. According to Graph A, the total production of GM and Ford companies is about _____ more than the combined production of Toyota, Renault, and Volkswagen.

   a. 500,000   **b.** 1,000,000   **c.** $\frac{1}{3}$ **d.** 2,000,000   **e.** 800,000

4. Which statement best describes Graph A?

   a. Toyota is catching up with Ford in car production.
   b. Volkswagen has had a bad year.
   c. In 1998 GM and Ford were the top producing cars.
   d. Americans like to drive American cars.
   e. 31,300,000 cars are produced every year.

GRAPH B

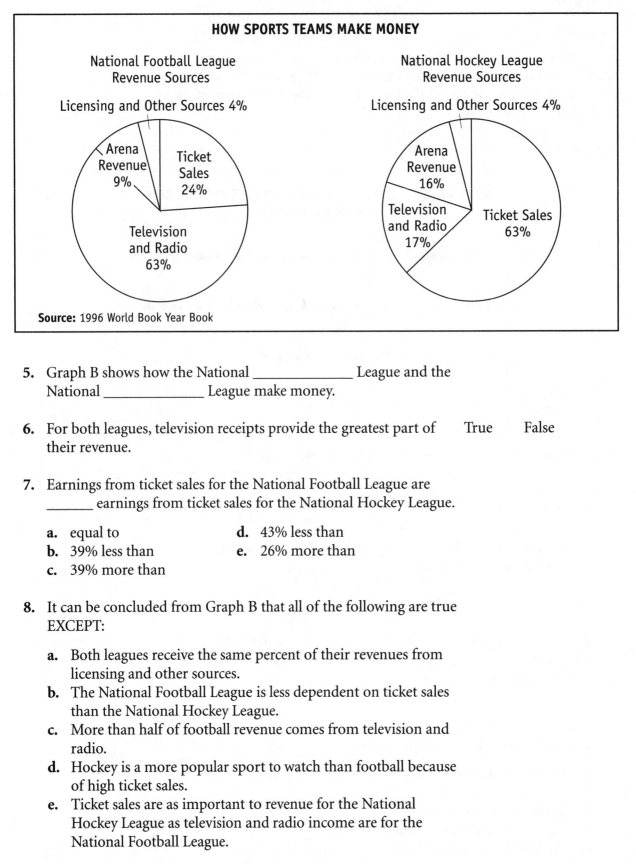

HOW SPORTS TEAMS MAKE MONEY

National Football League
Revenue Sources

Licensing and Other Sources 4%

Arena Revenue 9%
Ticket Sales 24%
Television and Radio 63%

National Hockey League
Revenue Sources

Licensing and Other Sources 4%

Arena Revenue 16%
Television and Radio 17%
Ticket Sales 63%

Source: 1996 World Book Year Book

5. Graph B shows how the National _____ League and the National _____ League make money.

6. For both leagues, television receipts provide the greatest part of their revenue.   True   False

7. Earnings from ticket sales for the National Football League are _____ earnings from ticket sales for the National Hockey League.

   a. equal to
   b. 39% less than
   c. 39% more than
   d. 43% less than
   e. 26% more than

8. It can be concluded from Graph B that all of the following are true EXCEPT:

   a. Both leagues receive the same percent of their revenues from licensing and other sources.
   b. The National Football League is less dependent on ticket sales than the National Hockey League.
   c. More than half of football revenue comes from television and radio.
   d. Hockey is a more popular sport to watch than football because of high ticket sales.
   e. Ticket sales are as important to revenue for the National Hockey League as television and radio income are for the National Football League.

GRAPH C

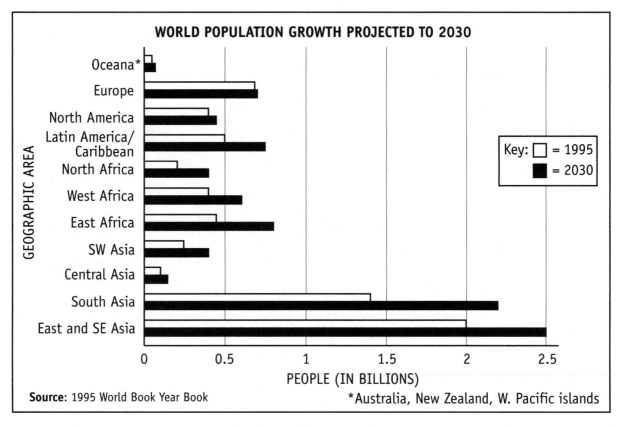

WORLD POPULATION GROWTH PROJECTED TO 2030

Key: □ = 1995
■ = 2030

GEOGRAPHIC AREA (vertical axis)

Oceana*
Europe
North America
Latin America/Caribbean
North Africa
West Africa
East Africa
SW Asia
Central Asia
South Asia
East and SE Asia

PEOPLE (IN BILLIONS)
0   0.5   1   1.5   2   2.5

Source: 1995 World Book Year Book   *Australia, New Zealand, W. Pacific islands

9. Graph C shows _____ in the world projected to
   _____.

10. The greatest population growth is projected to occur in Europe   True   False
    and North America.

11. What will be the increase in population in South Asia between 1995
    and 2030?

    a. $\frac{1}{2}$ billion
    b. 2 billion
    c. close to 1 billion
    d. less than $\frac{1}{2}$ billion
    e. more than 2 billion

12. Which statement best summarizes the information on Graph C?

    a. North American and East Africa will soon have the same
       population.
    b. The majority of the world's population in 2030 will live
       throughout Asia.
    c. Europe is losing population.
    d. The population of the world between 1995 and 2030 will
       probably double.

GRAPH D

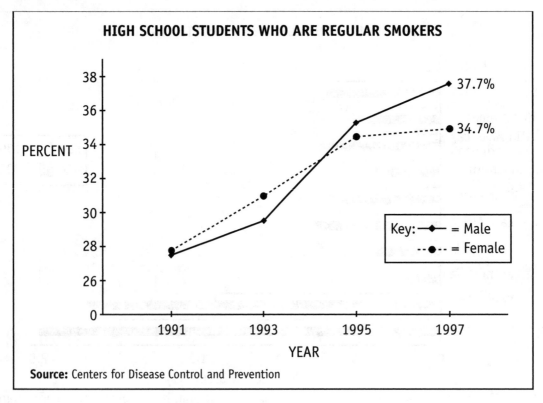

HIGH SCHOOL STUDENTS WHO ARE REGULAR SMOKERS

PERCENT

Source: Centers for Disease Control and Prevention

13. Graph D shows the increase in smoking for _____ from 1991 to _____.

14. In 1995 teenage girls were smoking as much as teenage boys.   True   False

15. In 1997 what was the difference in percentage of teenage boy smokers to teenage girl smokers?

   a. 3.7%
   b. 3.0%
   c. 4.0%
   d. 34.7%
   e. 37.7%

16. From the information on Graph D you could infer that

   a. the campaign against smoking is working
   b. teenagers are smoking at an earlier age
   c. teenage girls are smarter than teenage boys
   d. smoking is increasing rapidly among teenage girls
   e. smoking for teenage girls seems to be leveling off while it continues to rise for teenage boys

CHART A

AVERAGE UNEMPLOYMENT FOR 1998		
	Ages	
	16–19	20–24
Black men	30.1%	18.0%
Black women	25.3%	15.7%
White men	14.1%	6.7%
White women	10.9%	6.3%

Source: Bureau of Labor Statistics

17. Chart A shows percentages of _____ for 1998 by race.

18. By looking at Chart A, you can compare the difference in   True   False
    joblessness for people from teenage years to retirement.

19. The average unemployment for black men between ages 20 and
    24 is almost _____ that of white men of the same age.

    a. two times less
    b. two times greater
    c. three times less
    d. three times greater
    e. one-half greater

20. The following statement is true according to Chart A:

    a. More elderly people are unemployed than teenagers.
    b. Women have more trouble getting jobs than men.
    c. Black men consistently have higher unemployment rates than
       white men of the same age.
    d. The reason that white teenage women have low unemployment
       rates is because they stay in school.
    e. Unemployment is cut by half the older you get.

SCHEDULE A

City of New Orleans				
59		◀ Train Number ▶		**58**
Daily		◀ Days of Operation ▶		**Daily**
Read Down	▼		▲	**Read Up**
8 00P	Dp	Chicago, IL–Union Station	Ar	9 20A
8 51P		Homewood, IL		8 14A
9 24P		Kankakee, IL		7 39A
10 37P		Champaign-Urbana, IL		6 33A
11 18P		Mattoon, IL		5 44A
11 43P		Effingham, IL		5 17A
12 35A		Centralia, IL		4 28A
1 30A	Ar	Carbondale, IL	Dp	3 33A
1 35A	Dp		Ar	3 28A
3 40A		Fulton, KY		1 26A
4 25A		Newbern-Dyersburg, TN		12 36A
6 29A	Ar	Memphis, TN	Dp	10 53P
6 45A	Dp		Ar	10 41P
9 00A		Greenwood, MS		8 06P
9 59A		Yazoo City, MS		7 06P
11 09A		Jackson, MS		6 06P
11 48A		Hazlehurst, MS		5 14P
12 10P		Brookhaven, MS		4 53P
12 39P		McComb, MS		4 26P
1 45P		Hammond, LA		3 24P
3 30P	Ar	New Orleans, LA	Dp	2 15P

21. Schedule A shows the train schedule for Trains 59 and 58 which travel between _____ and _____.

22. The train called *City of New Orleans* travels through the states   True   False
    of Illinois, Kentucky, Tennessee, Mississippi, and Louisiana.

23. At what time does Train 59 going to New Orleans leave
    Greenwood, Mississippi?

    a. 9:00 P.M.
    b. 8:00 P.M.
    c. 8:00 A.M.
    d. 9:00 A.M.
    e. none of the above

24. If you left New Orleans at 2:15 in the afternoon, you would arrive
    in Chicago at

    a. 9:00 the next evening
    b. 9:20 the next evening
    c. 9:20 the same morning
    d. 9:20 the next morning
    e. 8:00 at night

MAP A

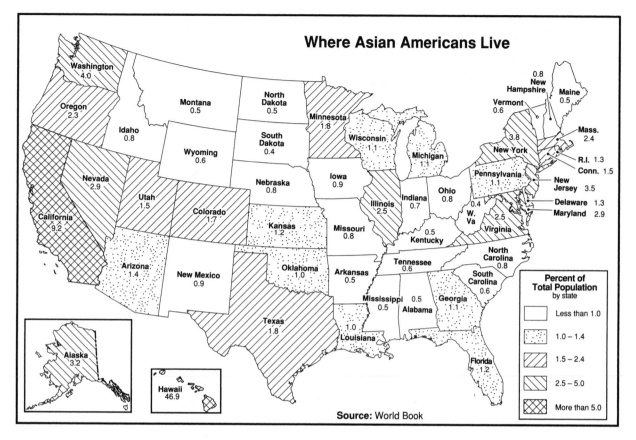

Where Asian Americans Live

Source: World Book

25. The map pictured above shows where _____ live in the United States.

26. According to the map, the states with the largest percentage   True   False
    of the Asian American population are California and Hawaii.

27. The percentage of Asian Americans in each state with the
    checkerboard design is _____ more than states not shaded at all.

    a. 3 times   **b.** $1\frac{1}{2}$ times   **c.** $2\frac{1}{2}$ times   **d.** 5 times   **e.** 4 times

28. Which statement best summarizes the information on the map?

    a. The six states in the north central part of the U.S. each have
       about the same number of Asian Americans living there.
    b. Most Asian Americans living in the U.S. are Chinese.
    c. The largest concentration of Asian Americans in the U.S. are in
       states bordering the Pacific and Atlantic Oceans.
    d. California has a larger percentage of Asian Americans in its
       population than New Mexico does.
    e. Asian Americans live in states where the climate is warm.

MAP B   **Downtown Pittsburgh**

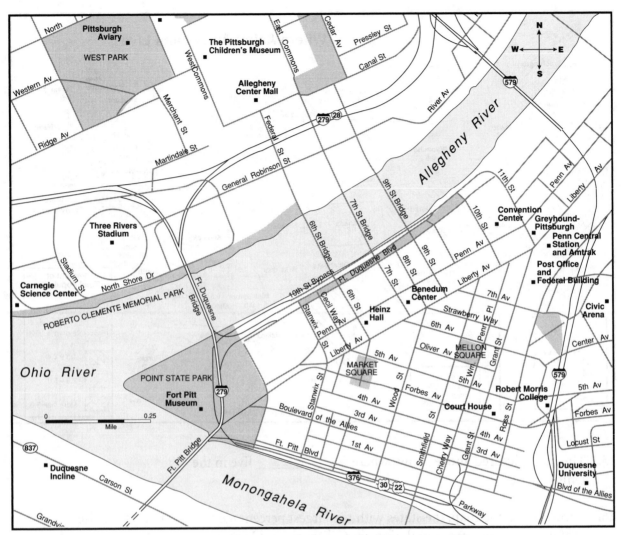

29. The rivers that join together in downtown Pittsburgh are the _____, _____, and _____.

30. It is more than a 5-mile drive between the Aviary in West Park   True   False
    and Fort Pitt Museum in Point State Park.

31. If you drive north on U.S. Interstate 579 from Duquesne University
    and go west on U.S. Interstate 279, then turn north on Federal
    Street, you will come to what point of interest?

    a. Civic Arena   **c.** Pittsburgh Aviary
    b. Three Rivers Stadium   **d.** Allegheny Center Mall

32. The directional map of downtown Pittsburgh contains the
    following information:

    a. location of airports   **c.** location of points of interest
    b. location of hotels   **d.** location of freeway exits

MAP C　　　　　　　　　　　**United States**

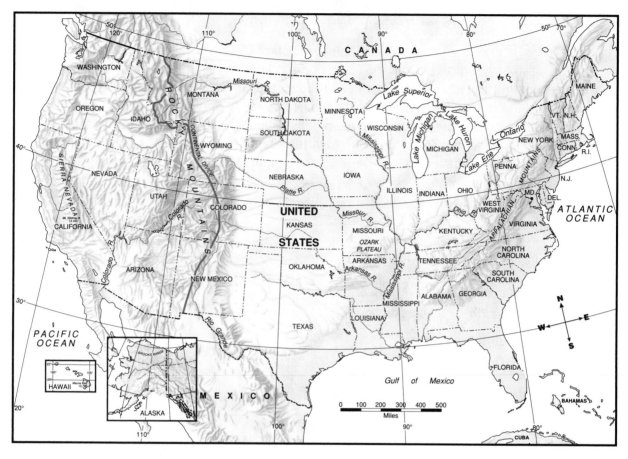

33. Map C shows the United States, which is bordered by the countries
_____ and _____.

34. The major mountain range in the eastern section of the　　　　True　　　False
United States is called the Sierra Nevada.

35. What are the two larger mountain ranges located west of the
Mississippi River?

　　a. Alaskan and Sierra Nevada　　　　　　　**d.** Sierra Nevada and Appalachian
　　b. Sierra Nevada and Ozark Plateau　　　　**e.** Rocky and Appalachian
　　c. Rocky and Sierra Nevada

36. It can be concluded from Map C that all of the following are true
EXCEPT:

　　a. Austin is the state capital of Texas.
　　b. Five great lakes lie between the northeast border of the United
　　　　States and Canada.
　　c. The eastern section of the United States possesses mountainous
　　　　areas.
　　d. The continental divide runs through the Rocky Mountains.
　　e. The United States has many rivers and mountains.

Pretest Chart

If you miss more than one problem in any section of this test, you should complete the lessons on the practice pages indicated on this chart. If you miss only one problem in a section of this test, you may not need further study in that chapter. However, before you skip those lessons, we recommend that you complete the review test at the end of that chapter. For example, if you miss one problem about graphs, you should pass the Graph Review (pages 68–75) before beginning the chapter on schedules and charts. This longer inventory will be a more precise indicator of your skill level.

PROBLEM NUMBERS	SKILL AREA	PRACTICE PAGES
1, 2, 3, 4	pictograph	20–31
5, 6, 7, 8	circle graph	32–43
9, 10, 11, 12	bar graph	44–55
13, 14, 15, 16	line graph	56–67
17, 18, 19, 20	chart	80–87, 92–95
21, 22, 23, 24	schedule	80–85, 88–95
25, 26, 27, 28	informational map	130–135
29, 30, 31, 32	directional map	124–129
33, 34, 35, 36	geographical map	118–123

Building
Number
Power

GRAPHS

Graph Skills Inventory

The Graph Skills Inventory allows you to measure your skills in reading and interpreting graphs. Correct answers are listed at the back of the book.

GRAPH A

LEADING PASSENGER AIRLINES
(approximate figures 1998)

Key: ✈ = 10 million passengers

Source: Air Transport Association of America, Inc.

Answer each question below.

1. The two airlines that carry the greatest number of passengers are United and _____.

2. Delta Airlines carries approximately _____ passengers each year.

3. Graph A represents the number of passengers carried on the six leading passenger airlines in 1998.   True   False

4. United Airlines and American Airlines carry approximately the same number of passengers every year.   True   False

5. What was the approximate total number of passengers carried by the six leading airlines in 1998?
    a. 400 million
    b. 355 million
    c. 440 million
    d. 280 million
    e. 180 million

6. Delta carried _____ as many passengers as Northwest.
    a. one-half
    b. twice
    c. one-fourth
    d. four times
    e. several times

GRAPH B

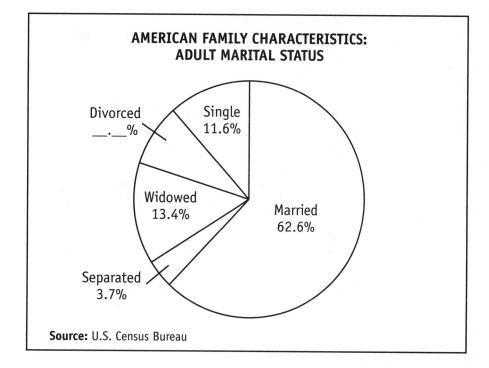

**AMERICAN FAMILY CHARACTERISTICS:
ADULT MARITAL STATUS**

Divorced __.__%

Single 11.6%

Widowed 13.4%

Married 62.6%

Separated 3.7%

Source: U.S. Census Bureau

Answer each question by filling in the blank, answering true or false, or by choosing the best multiple-choice response.

7. The categories of marital status shown are married, separated, _____ , divorced, and _____ .

8. The percent of the American population who are divorced is _____ . (Fill this in on the graph.)

9. More adults are separated and divorced than are single.   True   False

10. Three-quarters of adult Americans are married.   True   False

11. What is the percent of adult Americans who are not single?

   a. 88.4%   d. 50%
   b. 11.6%   e. 8.7%
   c. 91.3%

12. More than half of all adult Americans are

   a. single or widowed   d. separated or single
   b. divorced or widowed   e. widowed or separated
   c. married

GRAPH C

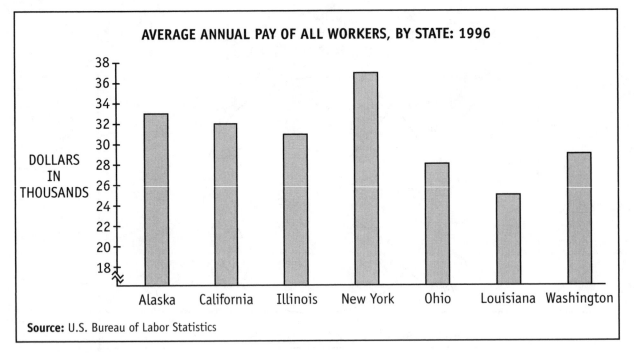

AVERAGE ANNUAL PAY OF ALL WORKERS, BY STATE: 1996

DOLLARS IN THOUSANDS

Alaska   California   Illinois   New York   Ohio   Louisiana   Washington

Source: U.S. Bureau of Labor Statistics

Answer each question by filling in the blank, answering true or false, or choosing the best multiple-choice response.

13. Graph C shows the average annual pay for all workers in different

    _____.

14. The average annual pay in Illinois was _____.

15. Average annual pay for all workers is shown for the states of   True   False
    New York and Florida.

16. The average annual pay in California was $32,000.   True   False

17. The average annual pay in Ohio is _____ than the average annual
    pay in Alaska.

    a. $8,000 more   **d.** $4,000 less
    b. $5,000 less   **e.** $4,000 more
    c. $6,000 less

18. Approximately what is the average annual pay for all workers in the
    seven states combined?

    a. $22,400   **d.** $39,900
    b. $30,700   **e.** $28,000
    c. $25,400

GRAPH D

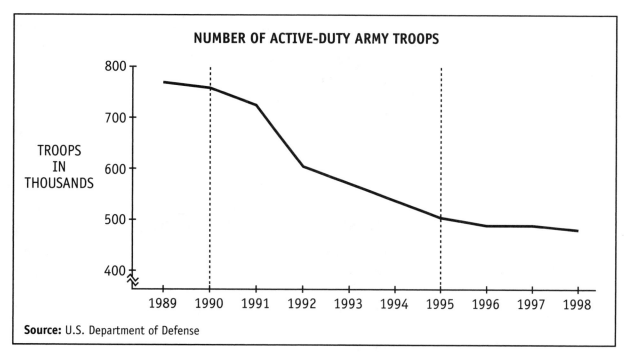

NUMBER OF ACTIVE-DUTY ARMY TROOPS

TROOPS IN THOUSANDS

Source: U.S. Department of Defense

Answer each question below.

19. In 1992 approximately _____ thousand troops were active in the U.S. Army.

20. From 1992 to 1998, there was a decrease of about 100,000 troops.   True   False

21. Between which years was the smallest decrease of Army troops?

    a. 1990–1994   **b.** 1995–1998   **c.** 1992–1995

Graph Skills Inventory Chart

Use this inventory to see what you already know about graphs and what you need to work on. A passing score is 18 correct answers. Even if you have a passing score, circle the number of any problem that you miss and turn to the practice pages indicated for further instruction.

PROBLEM NUMBERS	SKILL AREA	PRACTICE PAGES
1, 2, 3, 4, 5, 6	pictograph	16–31
7, 8, 9, 10, 11, 12	circle graph	16–19, 32–43
13, 14, 15, 16, 17, 18	bar graph	16–19, 44–55
19, 20, 21	line graph	16–19, 56–67

What Are Graphs?

A **graph** is a pictorial display of information. Since it is drawn, rather than written, a graph makes it possible to get a quick impression of a great deal of data and to easily make comparisons and draw conclusions. Graphs are often used in government, business, and education and may appear in reports, newspapers, and magazines.

Types of Graphs

In this workbook, you will study four main types of graphs.

PICTOGRAPHS

A **pictograph** uses pictures or symbols to display information.

A pictograph usually has a **key** to show the value of each symbol.

Pictographs are read by counting the **symbols** on a line of a graph and computing their value.

U.S. POPULATION BY STATES: 1997

Connecticut (Northeast)	🧍🧍🧍
Indiana (Midwest)	🧍🧍🧍🧍🧍🧍
Louisiana (South)	🧍🧍🧍🧍
Arizona (West)	🧍🧍🧍🧍🧍

Key: 🧍 = 1 million

Source: U.S. Census Bureau

CIRCLE GRAPHS

A **circle graph** uses parts of a circle to show information.

Circle graphs show values in each part of a divided circle. A part of a circle graph is called a **segment** or a **section.**

The segments of a circle add up to a whole or to 100% of the topic.

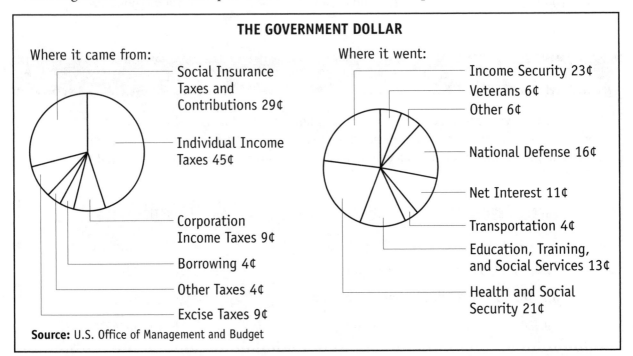

THE GOVERNMENT DOLLAR

Where it came from:
- Social Insurance Taxes and Contributions 29¢
- Individual Income Taxes 45¢
- Corporation Income Taxes 9¢
- Borrowing 4¢
- Other Taxes 4¢
- Excise Taxes 9¢

Where it went:
- Income Security 23¢
- Veterans 6¢
- Other 6¢
- National Defense 16¢
- Net Interest 11¢
- Transportation 4¢
- Education, Training, and Social Services 13¢
- Health and Social Security 21¢

Source: U.S. Office of Management and Budget

BAR GRAPHS

A **bar graph** uses thick bars to show information.

Bar graphs are usually drawn in one of two different directions:

1. With the bars running up and down. The bars are placed at equal distances along the **horizontal axis** that runs across the bottom of the graph.
2. With the bars running from left to right. The bars are placed at equal distances along the **vertical axis** on the left side of the graph.

To measure a bar accurately, you may want to lay the edge of a sheet of paper across the top of the bar to align with its value at the left (or at the bottom) of the graph.

Sometimes, a graph may have a break in the vertical axis and an open space running across the graph. This means that some values have been left off to save space on the graph.

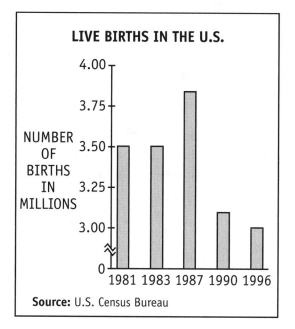

Source: U.S. Census Bureau

LINE GRAPHS

A **line graph** is drawn with one or more thin lines that extend across the graph.

Like the bar graph, a line graph is drawn using values along a horizontal and a vertical axis. Using the edge of a paper will help you measure locations on the graph accurately.

A line graph is most useful in showing trends and developments.

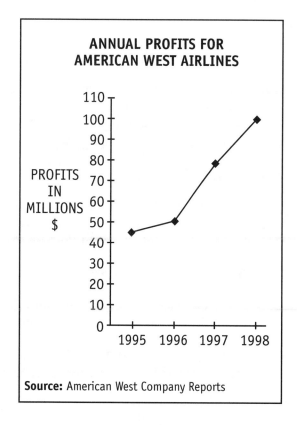

Source: American West Company Reports

Types of Questions

In this book, you are asked three types of questions. These questions will help you to find information and interpret graphs. Similar questions will be used in the sections on schedules, charts, and maps.

Scanning the Graph Questions

"Scanning the Graph" questions require you to look carefully at the graph and to fill in missing words to complete a sentence. To complete these statements, pay attention to the

- title of the graph
- names of axes or sections (of a circle graph)
- information in the key, if used
- source of the graph's information, if given

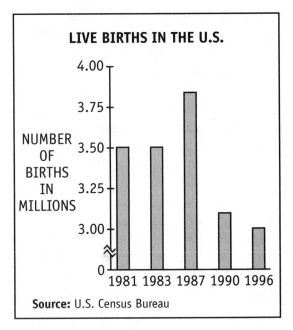

LIVE BIRTHS IN THE U.S.

Source: U.S. Census Bureau

EXAMPLE This graph tells about _____ in the United States.

ANSWER: live births. The title of the graph gives this information.

Reading the Graph Questions

"Reading the Graph" questions require you to determine specific information from the graph. You should respond either true or false to the given statements.

EXAMPLE There were more than 3 million live births in 1990.   True   False

ANSWER: True. Starting at the bottom of the graph (horizontal axis), find 1990. Then look at the top of the 1990 bar and follow a line across to a point on the vertical axis to the left. You can see that the 1990 bar goes higher than 3.00.

Comprehension Questions

Comprehension questions require you to do computations, make inferences, draw conclusions, or make predictions based on the graph's information. You will circle the letter of the best choice.

EXAMPLE What two years had approximately the same number of births?

    a. 1990 and 1996   **c.** 1983 and 1990   **e.** 1987 and 1996
    b. 1981 and 1983   **d.** 1987 and 1990

ANSWER: b. 1981 and 1983. Compare the heights of the bars. These bars are closer to the same height than any other two bars.

Read Carefully to Avoid Mistakes

Since interpretation is very important when answering questions, be sure to read a question carefully before choosing your response. Some questions may be tricky. The categories below are possible problems.

Information Is True but Not Contained on the Graph

EXAMPLE   According to Graph A, the profits for American West reflect   True   False
U.S. prosperity.

Answer: **False.** While this fact may be true, the graph gives you no information about our country's prosperity.

Misleading Words Are Given in Answer Choices

EXAMPLE   Which statement is true according to Graph A?

a. American West Airlines had the greatest profits in 1997.
b. Profits have risen each year.
c. Profits have ranged from $40 to $120 million.

ANSWER: **b.** The graph shows a line representing profits in millions of dollars increasing steadily from 1995 to 1998.

Answers **a.** and **c.** are not true; 1998 shows the greatest profits, and profits have ranged from $40 to $100 million, not $120 million.

GRAPH A

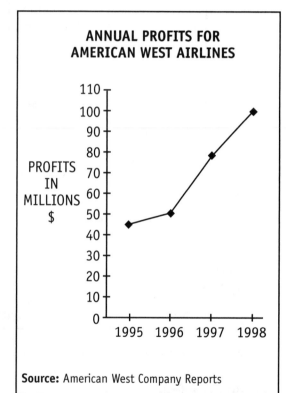

ANNUAL PROFITS FOR AMERICAN WEST AIRLINES

PROFITS IN MILLIONS $

Source: American West Company Reports

Labels Are Important

EXAMPLE   In 1997 approximately what were the average profits earned by American West Airlines?

a. $80 million
b. $72 million
c. $85 million
d. $50 million
e. $78 million

ANSWER: **e. $78 million.** Find 1997 on the horizontal axis. Follow up to the line and look across to the vertical axis to the left. You'll see a value of slightly below 80, or approximately $78 million.

Pictographs

A **pictograph** gets its name from the small pictures it uses as symbols on the graph. Pictographs generally use a **key** to show the value of the pictures that are used as symbols. Parts of symbols are often used to represent a fractional amount of a quantity shown in the key.

Pictographs are often not as exact as other types of graphs, but they are the easiest to read.

GRAPH A

PRINCIPAL U.S. LABOR UNIONS ← Title

Key: 👤 = 250,000 workers — Key

International Brotherhood of Electrical Workers	👤👤👤👤
United Food and Commercial Workers	👤👤👤👤👤
Communication Workers of America	👤👤
American Federation of Teachers	👤👤 ← Information Symbols
International Assoc. of Machinists & Aerospace Workers	👤👤👤
International Brotherhood of Teamsters	👤👤👤👤👤👤👤

Vertical Axis Names

Source: U.S. Department of Labor, Bureau of Labor Statistics

To answer questions about a pictograph, follow the sequence below.

Scanning the Graph

To scan a pictograph, find the graph title, the vertical axis names, and the key.

EXAMPLE   The Communication Workers of America is one of the principal _____.

STEP 1   Find the name "Communication Workers of America" on the vertical axis.

STEP 2   Look at the graph title: "Principal U.S. Labor Unions." This title gives the subject of the graph.

ANSWER: U.S. labor unions

Reading the Graph

To read a pictograph, count the number of symbols on a line. Then multiply the number of symbols by the value of the symbol given in the key.

Sometimes, only a part of the symbol is shown. Look at the partial symbol carefully. Most often, a partial symbol will be $\frac{1}{2}$ of the whole or sometimes $\frac{1}{4}$ or $\frac{3}{4}$. To find a value for a part of a symbol, find that fraction of the whole.

EXAMPLE The total membership of workers in the International   True   False
Brotherhood of Teamsters is 1,875,000.

STEP 1 Find the name "International Brotherhood of Teamsters."
STEP 2 Count the number of complete symbols: seven (7). Next, determine the fraction that the partial symbol represents: one-half symbol ($\frac{1}{2}$). There are $7\frac{1}{2}$ symbols.
STEP 3 Compute the value of the symbols. 1. Multiply the whole numbers. 2. Multiply the fraction. 3. Add.
1. $250,000 \times 7 = 1,750,000$
2. $250,000 \times \frac{1}{2} = +\ 125,000$
3. $\qquad\qquad\qquad\quad 1,875,000$

ANSWER: True

Comprehension Questions

Values on the graph can be compared and conclusions can be drawn.

EXAMPLE Between the American Federation of Teachers (AFT) and the International Brotherhood of Electrical Workers (IBEW), what is the difference in the number of members?

a. 250,000 workers   **d.** 500,000 workers
b. 400,000 workers   **e.** 100,000 workers
c. 750,000 workers

STEP 1 Find the name "American Federation of Teachers." Calculate this union's membership total by multiplying the number of symbols by the key's value.
$2 \times 250,000 = 500,000$
STEP 2 Find the name "International Brotherhood of Electrical Workers." Calculate this union's membership by multiplying the number of symbols by the key's value.
$4 \times 250,000 = 1,000,000$
STEP 3 Subtract to find the difference between the memberships of the two unions.
$1,000,000 - 500,000 = 500,000$

ANSWER: d. 500,000 workers

Practice Graph I

Graph I, read from left to right across the page, is an example of a horizontal pictograph. Graph I shows the hourly earnings of jobs in particular goods-producing and service-producing industries during 1998.

GRAPH I

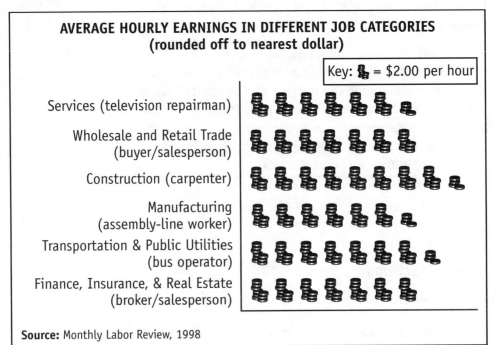

Source: Monthly Labor Review, 1998

Scanning Graph I

Fill in each blank as indicated.

1. Find each of the following:

   a. Graph Title: _____

   b. Vertical Axis Names:

   _____

   _____

   _____

   _____

   _____

   _____

2. Each 🪙 symbol is equal to _____ in earnings per hour.

3. The hourly pay for a carpenter is found by looking at the job category called _____.

Reading Graph I

Decide whether each statement is true or false and circle your answer.

4. The average hourly pay for a construction worker is $14 per hour.　　　True　　False

5. A bus operator can expect to earn average hourly wages of approximately $15 per hour.　　　True　　False

6. Graph I shows that salespeople working in clothing stores usually earn an average of $8 per hour.　　　True　　False

Comprehension Questions

Answer each question by choosing the best multiple-choice response.

7. Hourly earnings in the construction trades are generally _____ other occupations.

    a. higher than
    b. lower than
    c. the same as
    d. twice that of
    e. none of the above

8. What is the difference between the hourly wages of an assembly line worker and the earnings of a carpenter?

    a. $1.50
    b. $2.00
    c. $3.00
    d. $4.00
    e. $5.00

9. Which job category shown in Graph I has the highest hourly earnings?

    a. retail trade
    b. manufacturing
    c. transportation
    d. construction
    e. finances

10. Which statement best describes Graph I?

    a. More physically difficult jobs pay higher hourly wages.
    b. Workers skilled in a trade or a craft make lower wages than workers skilled in management or supervision.
    c. Average hourly earnings for union workers range from $2 to $3 per hour higher than for non-union workers.
    d. Average hourly earnings for workers in the job categories shown range from $13 to $17 per hour.
    e. Average hourly earnings for non-union workers are increasing at a rate faster than earnings for union workers.

Practice Graph II

Graph II is a vertical pictograph. Symbols are shown in vertical columns. A vertical pictograph is often used to compare information about one item over a period of time. Graph II shows the change in the unemployment rate over several years.

GRAPH II

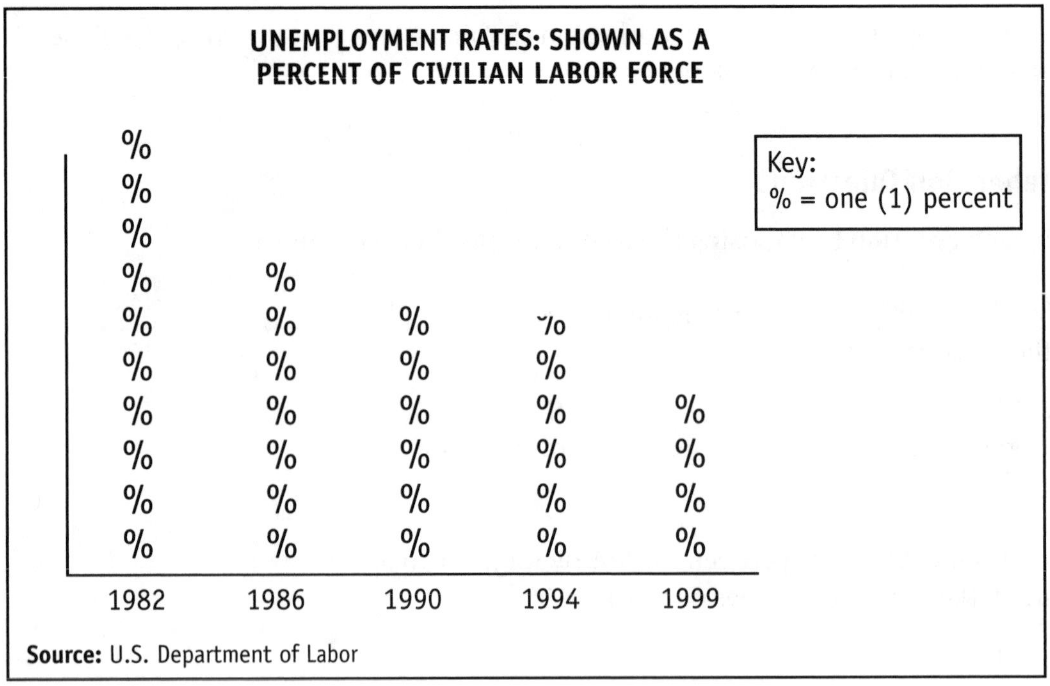

UNEMPLOYMENT RATES: SHOWN AS A PERCENT OF CIVILIAN LABOR FORCE

Key:
% = one (1) percent

1982	1986	1990	1994	1999

Source: U.S. Department of Labor

Scanning Graph II

Fill in each blank as indicated.

1. The information presented on Graph II was obtained from the source known as the _____.

2. The unemployment rates of the labor force are given from the year _____ to the year _____.

3. Graph II tells about the unemployment rates as a _____ of the civilian labor force.

Reading Graph II

Decide whether each sentence is true or false and circle your answer.

4. The highest unemployment rate shown occurred in 1986.   True   False

5. In 1986, the unemployment rate was about 7%.   True   False

6. According to Graph II, the lowest unemployment rate occurred   True   False
   in 1994.

Comprehension Questions

Answer each question by choosing the best multiple-choice response.

7. What is the difference in the unemployment rate between
   1986 and 1990?

   a. 1%
   b. 2%
   c. 3%
   d. 4%
   e. 5%

8. The greatest unemployment rate decrease occurred between
   what years?

   a. 1982–1986
   b. 1986–1990
   c. 1990–1994
   d. 1994–1999

9. From Graph II, you could assume that the greatest percentage of
   the workforce was employed in what year?

   a. 1999
   b. 1994
   c. 1990
   d. 1986
   e. 1982

10. Which statement best describes Graph II?

    a. Unemployment rates rose steadily from 1986 to 1990.
    b. Unemployment rates dropped steadily from 1982 to 1990.
    c. Unemployment rates continuously dropped from 1986 to 1999.
    d. Unemployment rates follow a consistent pattern of increasing
       one year and then decreasing the next.
    e. Unemployment rates were higher in 1994 than in any other
       year since 1982.

Practice Graph III

For the purpose of making comparisons, some pictographs, such as Graph III below, are used to display more than one type of information.

GRAPH III

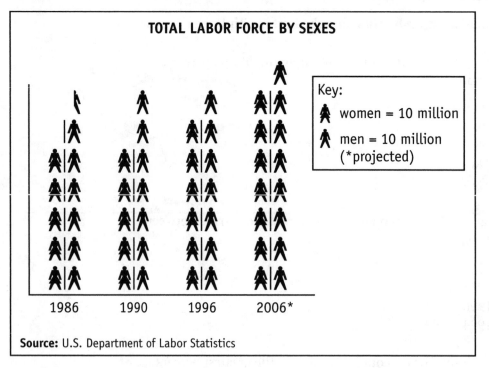

Source: U.S. Department of Labor Statistics

Scanning Graph III

Fill in each blank as indicated.

1. In Graph III, the asterisk (*) means that the information for 2006 is _____.

2. Each ♦ or ♦ on the graph represents _____ workers.
   (number)

3. Graph III is a labor force comparison between _____ and _____.

Reading Graph III

Decide whether each sentence is true or false and circle your answer.

4. The year with the greatest number of women in the work force was 1996. True False

5. During 1986 women made up more than one-half of the *total*   True   False
labor force.

6. By 1990 the *total* labor force was approximately 90 million   True   False
people.

Comprehension Questions

Answer each question by choosing the best multiple-choice response.

7. In 1996 the number of women in the work force was almost
_____ the number of men.

    a. equal to   **d.** four times
    b. twice   **e.** one-half
    c. triple

8. Between 1990 and 2006, the female labor force is projected to
increase by what amount?

    a. 85 million   **d.** 100 million
    b. 55 million   **e.** 10 million
    c. 20 million

9. Which diagram best represents a relationship shown on Graph III?

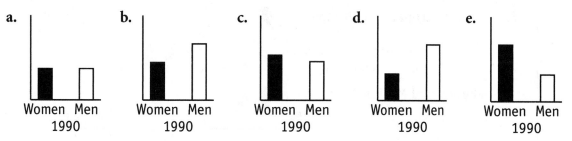

10. Which statement is true according to Graph III?

    a. In comparison to 1990, the number of people in the total labor
force is expected to decline in the 21st century.
    b. Men have been and are expected to continue as the larger of the
two groups of employed workers.
    c. Women are expected to become the larger group of employed
workers by the first part of the 21st century.

Pictographs: Applying Your Skills

When workers are unemployed, their families undergo many hardships. Some people may be able to find other work when they are fired or laid off. Others may not be so fortunate and must rely on unemployment compensation and public assistance for a while. One way to guard against becoming unemployed is to develop skills that will be in demand in the future and continually seek training and education.

GRAPH A

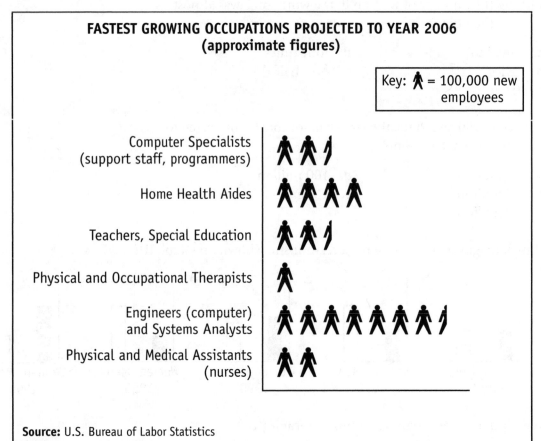

FASTEST GROWING OCCUPATIONS PROJECTED TO YEAR 2006
(approximate figures)

Key: 👤 = 100,000 new employees

Computer Specialists (support staff, programmers)	👤 👤 ⸝
Home Health Aides	👤 👤 👤 👤
Teachers, Special Education	👤 👤 ⸝
Physical and Occupational Therapists	👤
Engineers (computer) and Systems Analysts	👤 👤 👤 👤 👤 👤 👤 ⸝
Physical and Medical Assistants (nurses)	👤 👤

Source: U.S. Bureau of Labor Statistics

Answer each question by filling in the blank, answering true or false, or choosing the best multiple-choice response.

1. Graph A shows the fastest growing occupations to the year _____.

2. The total number of people projected to be employed as physical and medical assistants by 2006 will be approximately _____.

3. The number of home health aides is expected to be twice the number of physical and medical assistants.   True   False

4. According to Graph A, the number of teachers in 2006 will be 200,000.   True   False

5. The information shown in Graph A would indicate that more people are projected to be employed as home health aides than as physical therapists and computer specialists combined.   True   False

6. What is the approximate number of people projected for employment as engineers?

   a. over 600,000   d. below 400,000
   b. below 700,000   e. above 650,000
   c. over 700,000

7. According to Graph A, the difference in the number of people to be employed as computer specialists and _____ will be approximately $\frac{1}{2}$ million.

   a. health aides   d. physical and occupational therapists
   b. medical assistants e. engineers and system analysts
   c. teachers

8. If the number of people employed as physical and medical assistants increases by 50% more than the graph predicts, what will be the total number of assistants in 2006?

   a. 100,000   d. 400,000
   b. 200,000   e. 500,000
   c. 300,000

9. Which diagram best represents a relationship shown on Graph A?

   a.   b.   c.   d.   e.

   Engineers Teachers   Nurses Teachers   Health   Computer   Nurses Computer   Therapists Nurses
                                          Aides   Specialists   Specialists

10. All of the following conclusions can be drawn from Graph A EXCEPT:

    a. By the year 2006, more people will be employed as computer specialists than as physical and occupational therapists.
    b. By the year 2006, the number of teachers will increase by two times more than nurses.
    c. By the year 2006, there will be approximately 2 million new jobs in all of the computer occupations.

In an effort to establish fair working conditions for all, employees joined together to form unions in the mid-1800's. This alliance enabled the common worker to have a say in how industries were run.

Since their inception, unions have grown to incorporate many fields of labor. Workers join with others of their profession, enabling them to address issues important to their specific industry. The UAW has a much different agenda than the teacher's union, for example.

GRAPH B

U.S. MEMBERSHIP IN SOME AFL-CIO AFFILIATED UNIONS: 1999

Key: ☐ = 50,000 members

Plumbers, Pipefitters	☐☐☐☐☐☐
Painters and Allied Professionals	☐☐☐
Airplane Pilots	☐☐
Government Employees	☐☐☐☐☐☐☐☐☐☐
Electrical Workers	☐☐☐

Source: Industrial Relations Data and Information Services

Answer each question by filling in the blank, answering true or false, or choosing the best multiple-choice response.

1. Graph B represents the U.S. _____ in some AFL-CIO affiliated unions.

2. The union with the smallest membership as represented on Graph B is _____.

3. There are twice as many members in the electrical workers'   True   False
   union as in the airplane pilots' union.

4. The painters' union has a larger membership than the electrical   True   False
workers' union.

5. Graph B shows that of all the unions listed, government   True   False
employees have the largest membership.

6. Which unions, when combined, have about 200,000 members?

   a. plumbers and pilots
   b. government workers, pilots, and plumbers
   c. painters and pilots
   d. painters, electrical workers, and plumbers
   e. electrical workers and pilots

7. If the number of members in the plumbers and pipefitters' union
increased by 10%, how many members would this union have?

   a. 330,000   **d.** 270,000
   b. 410,000   **e.** 40,000
   c. 303,000

8. If the number of members in the _____ union were to decrease by
50%, this union would have only 150,000 members.

   a. government employees   **d.** pilots
   b. electrical workers   **e.** painters
   c. plumbers and pipefitters

9. From Graph B, how many more members are in the plumbers'
union than in the electrical workers' union?

   a. 100,000   **d.** 600,000
   b. 400,000   **e.** 300,000
   c. 150,000

10. Which statement is true according to Graph B?

   a. There are four times more members in the government
   employees' union than in the electrical workers' union.
   b. There are more than twice as many members in the
   government employees' union than in the plumbers and
   pipefitters' union.
   c. 50% of all electrical workers belong to a union.
   d. In 1999 there were more members in AFL-CIO affiliated unions
   than ever before.
   e. Membership in unions is declining overall.

Circle Graphs

A **circle graph,** or **pie graph,** shows an entire quantity divided into various parts. Each part of the circle is called a **segment** and has its own name and value. In most cases, values on circle graphs consist of either parts, a fraction, or a percent of a whole.

Circle graphs often illustrate budgets and expenses. Graph A shows the sources for each dollar that the federal government receives.

GRAPH A

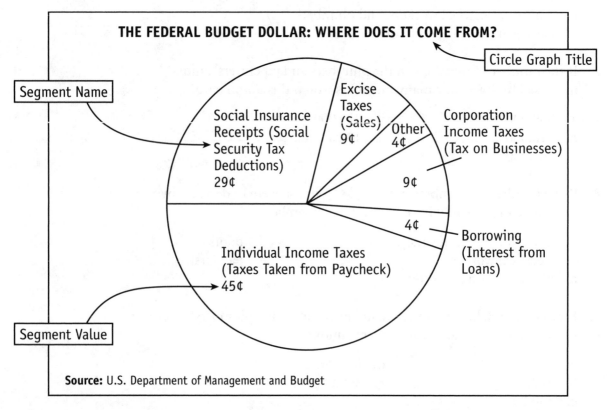

To answer questions about a circle graph, follow the directions below.

Scanning the Graph

To scan a circle graph, read the graph title and the segment names.

EXAMPLE   Graph A shows that part of the federal budget dollar comes from individual income _____.

STEP 1   Find and read the graph title, "The Federal Budget Dollar: Where Does It Come From?" The title is important but does not give the answer.

STEP 2   Read the segment names. Read clockwise until you find the segment name, "Individual Income Taxes."

ANSWER: taxes

Reading the Graph

To read a circle graph, locate the name and value of each segment.
Each segment value can represent a fraction, percent, or number.

EXAMPLE   From every dollar the government receives, 29¢ comes   True   False
            from social insurance receipts.

> **STEP 1**   Read clockwise on the graph until you find the segment
>   named "Social Insurance Receipts."

> **STEP 2**   Near the segment name, read the dollar value. The value
>   for social insurance receipts is 29¢.

ANSWER: True

Comprehension Questions

By comparing segment names and segment values, you can draw
conclusions from the graph.

EXAMPLE   For each dollar, what is the difference between the amount
            that the government receives from excise taxes and from
            borrowing?

> **a.** 1¢
> **b.** 5¢
> **c.** 10¢
> **d.** 25¢
> **e.** 50¢

> **STEP 1**   Find the segment name and value "Excise Taxes—9¢."

> **STEP 2**   Find the segment name and value "Borrowing—4¢."

> **STEP 3**   Find the difference between the two values by
>   subtracting.

> $$\begin{array}{r} 9\text{¢} \\ -\ 4\text{¢} \\ \hline 5\text{¢} \end{array}$$

ANSWER: b. 5¢

Practice Graph I

Graph I is a circle graph that shows where each dollar budgeted by the federal government is spent. Each pie-shaped segment represents a part of a dollar ($1.00).

GRAPH I

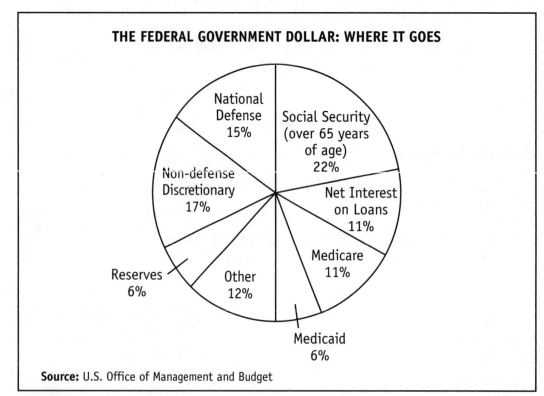

THE FEDERAL GOVERNMENT DOLLAR: WHERE IT GOES

National Defense 15%

Social Security (over 65 years of age) 22%

Non-defense Discretionary 17%

Net Interest on Loans 11%

Reserves 6%

Other 12%

Medicare 11%

Medicaid 6%

Source: U.S. Office of Management and Budget

Scanning Graph I

Fill in each blank as indicated.

1. Find each of the following:

   a. This graph is about _____.

   b. The main categories of the federal budget are

   _____   _____

   _____   _____

   _____   _____

   _____   _____

2. Payments made to citizens over 65 come from _____.

3. The category to which Congress allocates money for missiles, tanks, and military staff pay is _____.

Reading Graph I

Decide whether each statement is true or false and circle your answer.

4. At least one-half of each federally budgeted dollar is used for Medicaid and Medicare health payments.   True   False

5. Of each dollar the government spends, 25% is spent on national defense.   True   False

6. The smallest amount spent in any one category is in non-defense expenditures.   True   False

Comprehension Questions

Answer each question by choosing the best multiple-choice response.

7. From each dollar collected, the government spends _____ toward net interest on loans.

   a. 1 cent
   b. 6 cents
   c. 11 cents
   d. 17 cents
   e. 22 cents

8. The difference between money spent for national defense and money paid in Social Security benefits to individuals is _____ for each dollar.

   a. 14¢
   b. 6¢
   c. 11¢
   d. 8¢
   e. 7¢

9. Approximately, how much of the budget is spent on areas *other than* Social Security and Medicare?

   a. $\frac{1}{3}$
   b. $\frac{1}{4}$
   c. $\frac{2}{3}$
   d. $\frac{3}{4}$
   e. $\frac{4}{5}$

10. Which statement summarizes Graph I?

   a. The federal dollar is equally divided among several categories.
   b. The federal government collects most of the money for its budget from personal income taxes.
   c. The two leading government expenditures are for Social Security payments and non-defense discretionary.
   d. The interest paid on the national debt increased 10% between 1989 and 1990.

Practice Graph II

Graph II shows the number of deaths caused by motor vehicles in 1997. Motor vehicle crashes are the leading cause of death among Americans ages 1 to 34. Each segment of the circle represents the percentage of people killed in each category of motor vehicle accidents (or as a pedestrian killed by a motor vehicle).

GRAPH II

DEATHS CAUSED BY MOTOR VEHICLES, 1997

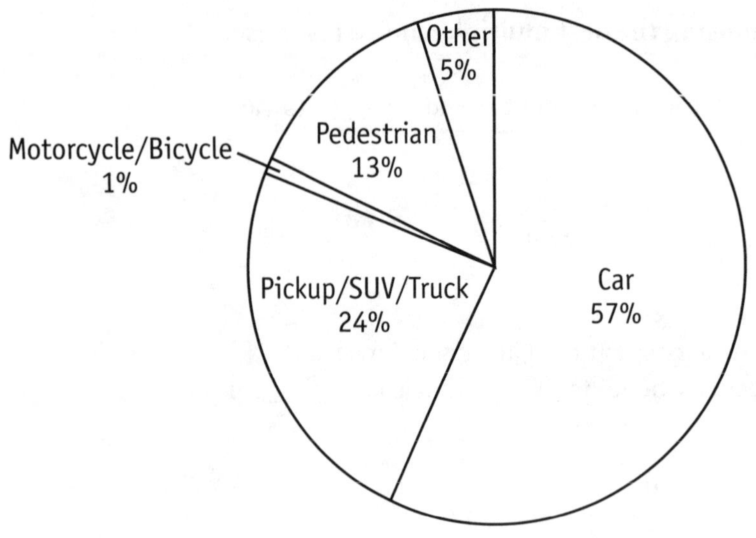

Other
5%

Motorcycle/Bicycle
1%

Pedestrian
13%

Pickup/SUV/Truck
24%

Car
57%

Source: Insurance Institute for Highway Safety, 1998

Scanning Graph II

Fill in each blank as indicated.

1. Graph II shows the percent of deaths caused by _____
   _____.

2. From Graph II, a person walking in a crosswalk who gets hit by a car would be included as a death in the category of _____.

3. The causes of death are divided among _____ categories.
   (number)

Reading Graph II

Decide whether each sentence is true or false and circle your answer.

4. 47% of all deaths in motor vehicles are caused by car accidents.   True   False

5. Sport utility vans and trucks show an increase in causes of deaths.   True   False

6. From Graph II, slightly more than three-fourths of all deaths are   True   False
   caused by trucks and cars.

Comprehension Questions

Answer each question by choosing the best multiple-choice response.

7. Deaths caused by vans/trucks exceed pedestrian and motorcycle
   deaths by _____ percent.

   a. 24   d. 10
   b. 8   e. 14
   c. 16

8. If the percentage of pedestrian deaths doubles in the next year,
   what will be the new allocation for deaths in this category?

   a. 8%   d. 5%
   b. 14%   e. 17%
   c. 26%

9. From Graph II, what causes approximately one-fourth of all motor
   vehicle deaths?

   a. Car   d. Other
   b. Pickup/SUV/Truck   e. Pedestrian
   c. Motorcycle/Bicycle

10. All of the following statements can be inferred from Graph II
    EXCEPT:

    a. More persons die in cars than any other vehicle type.
    b. The majority of deaths are caused by vans, trucks, and cars.
    c. Over half of deaths in motorized vehicles are due to car
       accidents.
    d. The majority of deaths to pedestrians are caused by a car
       or truck.
    e. Motorcycles and bike accidents total the fewest deaths caused
       by motor vehicles.

Practice Graph III

In some cases, more than one graph can be used to make comparisons. The following circle graphs show how the federal budget was estimated to change during three consecutive years.

GRAPH III

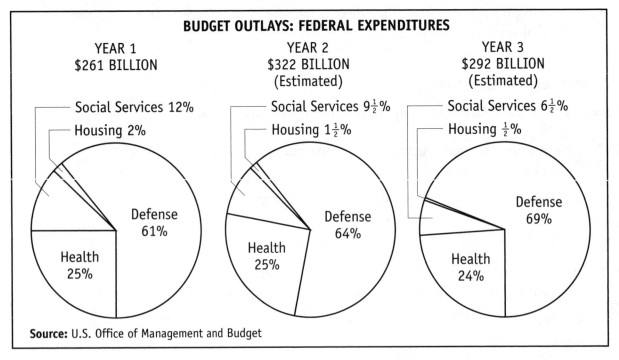

BUDGET OUTLAYS: FEDERAL EXPENDITURES

YEAR 1 — $261 BILLION

YEAR 2 — $322 BILLION (Estimated)

YEAR 3 — $292 BILLION (Estimated)

Source: U.S. Office of Management and Budget

Scanning Graph III

Fill in each blank as indicated.

1. The amount of federal expenditures shown for Year 2 and Year 3 are _____ figures.

2. The total budget for Year 2 and Year 3 is estimated to be $_____ billion and $_____ billion, respectively.

3. According to Graph III, the federal government's budget is divided into these four main categories:

    _____

    _____

    _____

    _____

Reading Graph III

Decide whether each sentence is true or false and circle your answer.

4. In Year 1, 2% of federal expenditures was for housing.   True   False

5. The percent of the federal budget allocated for defense is expected to decrease over the 3-year period.   True   False

6. The budget outlay for defense in Year 3 is estimated to be 69%.   True   False

Comprehension Questions

Answer each question by choosing the best multiple-choice response.

7. From Year 1 to Year 3, the percent allocated for which item will change the least?

   a. health   **c.** social services
   b. housing   **d.** defense

8. What budget item was projected to be reduced more than any other by Year 3?

   a. defense   **c.** social services
   b. housing   **d.** health

9. Which diagram best shows the total budget outlays?

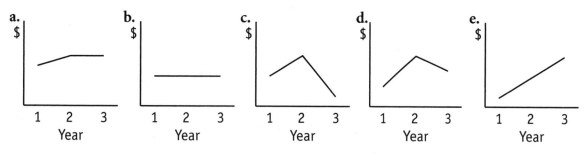

10. Which statement best reflects the information in Graph III?

   a. For Year 1 to Year 3, there is a projected increase of total federal expenditures in all categories.
   b. For Year 1 to Year 3, there is a projected increase in the percentage of total federal expenditures going for defense.
   c. For Year 2, total federal expenditures were projected to be less than they were in year 1 for both housing and health.
   d. Unemployment will increase in Year 3 because of decreasing federal expenditures for social services.

Circle Graphs: Applying Your Skills

When state and federal governments cut their budgets, the effects are felt by various agencies, organizations, and, most directly, by individual citizens. The following articles and graphs reveal interesting information.

---

3 MILLION FACE LOSS OF FOOD STAMPS

Budget cutbacks are placing severe burdens on low-income families. The poor and unemployed depend, to some extent, on the government to help provide food, clothing, and other essentials.

GRAPH A

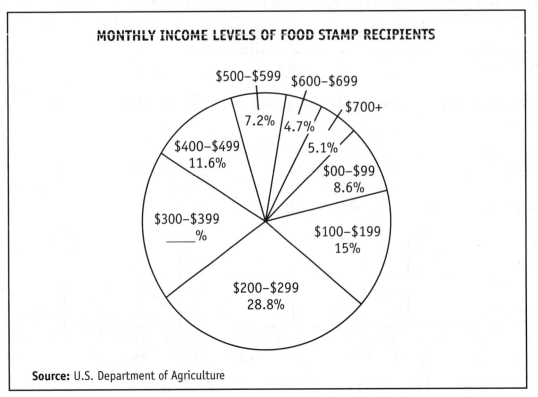

MONTHLY INCOME LEVELS OF FOOD STAMP RECIPIENTS

$500–$599 7.2%
$600–$699 4.7%
$700+
5.1%
$00–$99 8.6%
$400–$499 11.6%
$100–$199 15%
$300–$399 ____%
$200–$299 28.8%

Source: U.S. Department of Agriculture

Answer each question by filling in the blank, answering true or false, or choosing the best multiple-choice response.

1. Graph A shows the percentage of households receiving food stamps that are earning a specific _____.

2. The percent of food stamp recipients in the $300 to $399 category must be _____. (Fill in this amount on the graph.)

3. Households earning in excess of $700 each month do not qualify for food stamps.   True   False

4. Graph A shows that 15% of households receiving food stamps earn between $100–$199 per month.   True   False

5. The largest percentage of households that receive food stamps earn from $300–$399 per month.   True   False

6. 11.6% of all households receiving food stamps are within what monthly income range?

    a. $100–$199
    b. $200–$299
    c. $300–$399
    d. $400–$499
    e. $500–$599

7. What percent of all households that receive food stamps earn less than $400 each month?

    a. 83%
    b. 71.4%
    c. 19%
    d. 47.8%
    e. none of the above

8. Approximately twice as many households earning $200–$299 each month receive food stamps as households earning _____ each month.

    a. $100–$199
    b. $300–$399
    c. $400–$499
    d. $500–$599
    e. $600–$699

9. Which diagram best illustrates the relationship of percentages on Graph A?

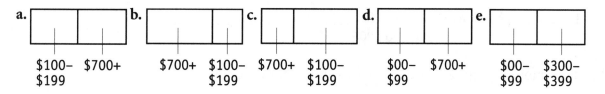

a. $100–$199 | $700+
b. $700+ | $100–$199
c. $700+ | $100–$199
d. $00–$99 | $700+
e. $00–$99 | $300–$399

10. Graph A is best summarized as follows:

    a. The largest percentage of households that receive food stamps earn monthly incomes between $200–$299.
    b. Households with incomes above $700 each month are no longer eligible for food stamps.
    c. Household income levels are the only factor in determining the amount of food stamps received.

Graph B shows two circle graphs used to compare the changes in civilian labor forces projected to the year of 2006.

Because of changes in work, education, and mobility, individuals are tending to move where the jobs are and where opportunity knocks. For many individuals and their families, the United States offers opportunities that don't exist in other parts of the world.

GRAPH B

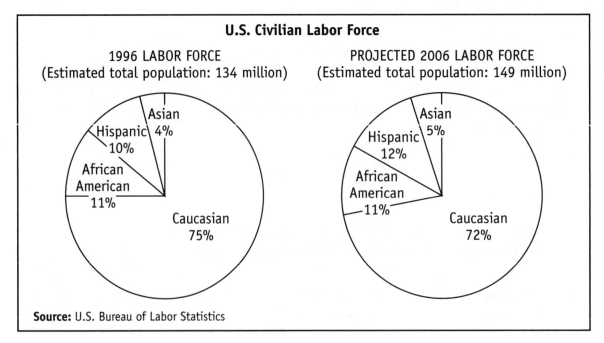

U.S. Civilian Labor Force

1996 LABOR FORCE
(Estimated total population: 134 million)

PROJECTED 2006 LABOR FORCE
(Estimated total population: 149 million)

Asian 4%
Hispanic 10%
African American 11%
Caucasian 75%

Asian 5%
Hispanic 12%
African American 11%
Caucasian 72%

Source: U.S. Bureau of Labor Statistics

Answer each question by filling in the blank, answering true or false, or choosing the best multiple-choice response.

1. The graphs above show the U.S. _____ _____
    _____ for 1996 and 2006.

2. A more appropriate title for the two graphs above could be

    a. U.S. Civilian and Army Labor Force
    b. U.S. Civilian Labor Force by Ethnic Origin
    c. U.S. Labor Force by Sex, Race, and Ethnicity
    d. Percentage of Changes in Total Labor Force

3. Graph B shows that the African American labor force   True   False
    is projected to remain the same.

4. There are about twice as many Hispanics projected in the U.S. labor force in 2006 as compared to 1996.   True   False

5. From the graphs, you can infer that more minorities are projected to enter the labor force.   True   False

6. The majority of the civilian labor force consists of which group?

   a. African Americans   c. Hispanics
   b. Caucasians   d. Asians

7. The minority civilian labor force is projected to increase from 25% in 1996 to _____% in 2006.

   a. 28   b. 23   c. 30   d. 35   e. 72

8. From 1996 to 2006, which group is projected to have the greatest increase in U.S. civilian labor force?

   a. African Americans   c. Asians
   b. Caucasians   d. Hispanics

9. Which diagram best compares the percent of changes in ethnic origin from 1996 to 2006?

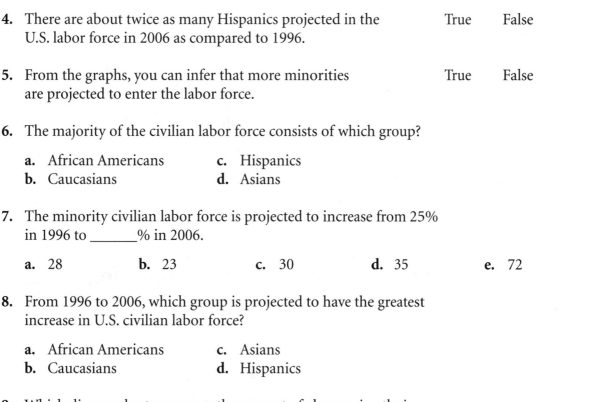

   a. African Americans   b. Hispanics   c. Asians   d. Hispanics   e. Caucasians

10. All the statements below can be concluded from Graph B EXCEPT:

   a. There is an increase in minorities across the U.S. labor force.
   b. The percent of minority workers is smaller in 1996 than projected in 2006.
   c. Caucasians comprise the largest percentage of the U.S. civilian labor force.
   d. More individuals are entering the labor force than in the past.
   e. Less than 25% of the U.S. civilian labor force in 2006 are minorities.

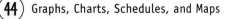

Bar Graphs

Bar graphs get their name from the thick bars with which they are drawn. They are generally more complicated to read than pictographs or circle graphs.

Bar graphs contain labeled points (usually names or numbers) along the **horizontal** and **vertical axes.** The values represented by an individual bar, called an **information bar,** are determined by both its height (or length) and by its position along an axis. Follow the steps outlined below to learn how to accurately read bar graphs.

GRAPH A

RECOMMENDED DAILY CALORIC INTAKE FOR AMERICAN BOYS

Bar Graph Title

CALORIES

3000
2700
2400
2100
1800
1500
1200
900
600
300
0

Vertical Axis Names

Information Bars

1–3   4–6   7–10   11–14

AGE GROUPS

Horizontal Axis Names

Source: National Academy of Sciences

To answer questions about bar graphs, follow the sequence below.

Scanning the Graph

To scan a bar graph, find the graph title and the names of the axes.

EXAMPLE   Graph A shows the recommended daily calorie intake for American boys whose ages range from 1 to _____ years.

STEP 1   Find the horizontal axis name Age Groups.

STEP 2   Scan from left to right, reading the labeled points at the bottom of each information bar. The bar farthest to the right represents the oldest age group, 11–14.

ANSWER: 14

Reading the Graph

To read a bar graph, find the values represented by the information bar. Read these values as labeled points along the horizontal and vertical axes.

EXAMPLE The amount of daily calories recommended for a boy 12 years of age is 2,700.

True   False

STEP 1 Find the horizontal axis name, "11–14" years old. This axis name includes those boys whose ages are 11, 12, 13, and 14.

STEP 2 Move your eyes from the bottom to the top of this information bar. After doing this, move across the graph to the left until you reach the vertical axis. You might use the edge of a paper to make sure you are aligning accurately.

STEP 3 Read the vertical axis name and number, "Calories, 2,700." The number of calories for a boy 12 years old is 2,700.

ANSWER: True

Comprehension Questions

Comprehension questions involve comparing values represented by several information bars and then drawing conclusions. Also, bar graphs are useful in showing trends. A **trend** is a pattern that can be seen from the information contained on the graph. From a trend, it is often possible to make predictions about future occurrences.

EXAMPLE As boys get older, the number of calories consumed daily should

a. stay the same   **c.** decrease
b. increase   **d.** level off

STEP 1 Find the horizontal axis names that refer to the specific age groups. Scan from the bottom to the top of each bar and identify the calories for each group.

STEP 2 Compare the amount of calories between the ages of boys from very young to older. For example, ages:

1–3 =	1,300 calories
4–6 =	1,700 calories
7–10 =	2,400 calories
11–14 =	2,700 calories

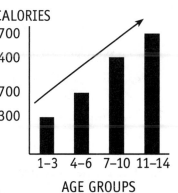

STEP 3 Draw a conclusion: As boys get older, the number of calories they require increases.

ANSWER b. increase

Practice Graph I

Graph I shows a vertical bar graph with bars drawn up from the horizontal axis. Also, notice the symbol ≈ on the vertical axis line. Occasionally, this symbol is used to show that values have been omitted from the axis in order to save space. On the graph below, pounds from 1 to 135 have been omitted.

GRAPH I

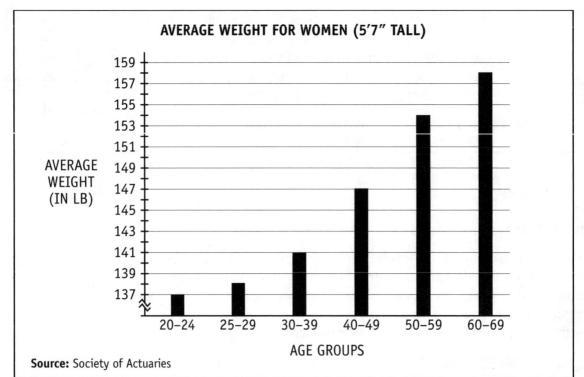

Source: Society of Actuaries

Scanning Graph I

Fill in each blank as indicated.

1. Find each of the following.

    a. Graph Title: _____

    b. Vertical Axis Title: _____

    c. Horizontal Axis Title: _____

2. The youngest women represented are from _____ to _____ years old.

3. The greatest average weight shown on the graph for women 5'7" tall is _____ pounds.

Reading Graph I

Decide whether each statement is true or false and circle your answer.

4. The average weight for a 40-year-old woman is 147 pounds.   True   False

5. A 152-pound woman in the age group 30–39 years is below the average weight.   True   False

6. The highest average weight occurs for women in the 50–59 year age group.   True   False

Comprehension Questions

Answer each question by choosing the best multiple-choice response.

7. On the average, as women get older, the trend is for their weight to

    a. decrease
    b. stay the same
    c. increase
    d. increase until age 30
    e. none of the above

8. What is the difference in average weight between a 69-year-old woman and a 29-year-old woman?

    a. 12 pounds
    b. 10 pounds
    c. 20 pounds
    d. 18 pounds
    e. 35 pounds

9. From Graph I, you can tell that the greatest gain in weight occurs from age group _____ to _____.

    a. 20–24, 25–29
    b. 25–29, 30–39
    c. 30–39, 40–49
    d. 40–49, 50–59
    e. 50–59, 60–69

10. Graph I tells about

    a. the average weight and height for 5-foot 7-inch-tall adults
    b. average weights for all women between the ages of 20 and 69 years
    c. average weight at specific ages for women 5 feet 7 inches tall
    d. average weights for obese women

Practice Graph II

Graph II is a horizontal bar graph. The information bars are drawn across the graph from left to right. Although they are not as common as vertical bar graphs, it is useful to know how to read horizontal bar graphs.

GRAPH II

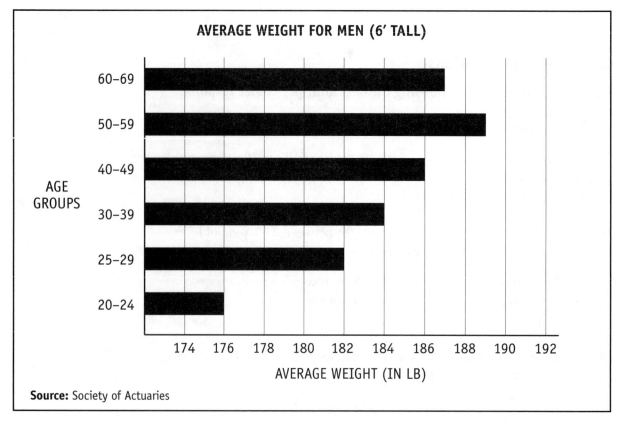

Source: Society of Actuaries

Scanning Graph II

Fill in each blank as indicated.

1. Graph II represents the _____ for men 6 feet tall.

2. The highest weight that could be represented on this graph is _____.

3. The total age span represented by Graph II starts at age _____ and extends to age _____.

Reading Graph II

Decide whether each sentence is true or false and circle your answer.

4. The average weight for a 6-foot-tall, 28-year-old man is   True   False
    184 pounds.

5. The graph shows that the lowest average weight for 6-foot-tall   True   False
    men occurs in the 20 to 24 year age group.

6. A 6-foot-tall man in the 50- to 59-year age group may weigh   True   False
    188 pounds.

Comprehension Questions

Answer each question by choosing the best multiple-choice response.

7. On the average, a man's weight can be expected to increase until
    what age?

    a. 45   **d.** 63
    b. 49   **e.** 60
    c. 39

8. The greatest average weight increase shown on the graph occurs
    from the ages of

    a. 20–24 to 25–29   **d.** 40–49 to 50–59
    b. 25–29 to 30–39   **e.** 50–59 to 60–69
    c. 30–39 to 40–49

9. Which graph best shows the weight trend for men 6 feet tall
    between the ages of 20 and 69?

10. Which statement best describes the information on Graph II?

    a. Men tend to gain weight as their age increases.
    b. On the average, men 6 feet tall gain weight up to age 59, and
            tend to lose weight after age 60.
    c. Taller men gain more weight in their later years than shorter men.
    d. Men tend to gain weight during retirement years.
    e. Weight gain is a problem for most people.

Practice Graph III

Graph III is a double bar graph that contains two kinds of information on the same graph. It shows the average weight for both men and women who are 5 feet 5 inches tall.

In each age group, a bar representing women's weight stands next to a bar representing men's weight. The bars are identified by a key in the upper right hand corner of the graph. Graphs containing more than one bar are used to make comparisons.

GRAPH III

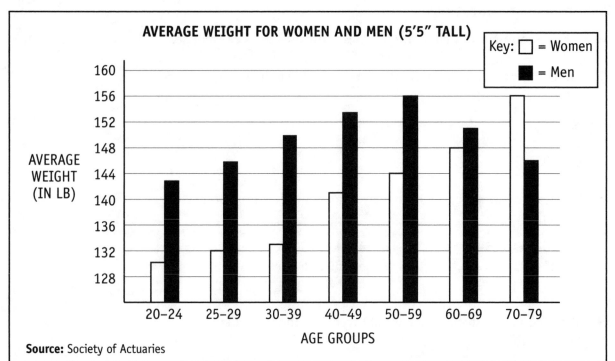

Source: Society of Actuaries

Scanning Graph III

Fill in each blank as indicated.

1. In Graph III, ■ represents the average weight for _____, and ☐ represents the average weight for _____.

2. The youngest age group represented on Graph III is from age _____ to _____, and the oldest age group is from _____ to _____.

3. The highest weight that could be represented on this graph is _____.

Reading Graph III

Decide whether each sentence is true or false and circle your answer.

4. At age 53, a 5'5" man is likely to weigh about 12 pounds more than a 5'5" woman of the same age.   True   False

5. When both are in their seventies, 5'5" men are usually heavier than 5'5" women.   True   False

6. The average weights of 5'5" men and women are the closest to being equal between ages 60 and 69 years.   True   False

Comprehension Questions

Answer each question by choosing the best multiple-choice response.

7. From 30 to 39 years, 5'5" women weigh about _____ pounds less than 5'5" men.

   a. 4
   b. 8
   c. 10
   d. 16
   e. 20

8. The trend on Graph III indicates that as a woman gets older, she is expected to _____ weight.

   a. gain
   b. lose
   c. gain and then lose
   d. lose and then gain

9. Women who are 5'5" tall and between the ages of 70–79 are the closest in weight to men 5'5" tall and between what ages?

   a. 20–24
   b. 25–29
   c. 40–49
   d. 50–59
   e. 60–69

10. Which statement best describes the information on Graph III?

    a. In the age span shown, the pattern of weight gain for men is the same as the pattern of weight gain for women.
    b. Through their fifties, taller men and women are heavier than shorter men and women.
    c. Both men's and women's weight are affected by retirement.
    d. Both men and women tend to gain weight through their fifties, but men, unlike women, tend to lose weight in later years.

Bar Graphs: Applying Your Skills

Americans are becoming more concerned with the quality of life and with their health. Weight and disease control, nutrition, and physical fitness are popular topics. Additionally, our society is placing great importance on improving health care services while controlling costs.

Graph A shows the sources of health care payments.

GRAPH A

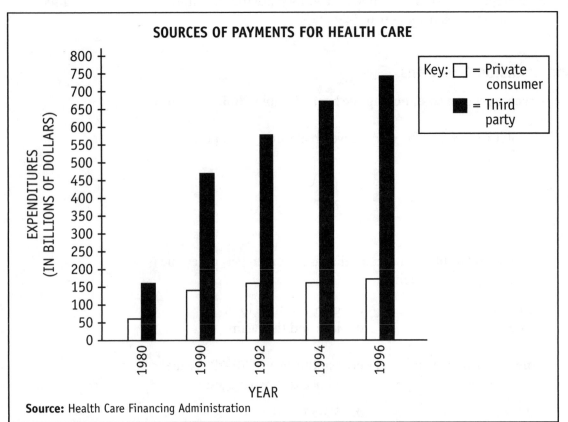

Answer each question by filling in the blank, answering true or false, or choosing the best multiple-choice response.

1. The sources of payments for health care shown on the graph are _____ and _____.

2. From 1980 to 1990, third party expenditures increased by approximately _____ dollars.

3. Private consumer payments were about half the amount of third party payments in 1980.   True   False

4. From 1992 to 1994, private consumer payments remained about the same amount.   True   False

5. In 1994, third party health care payments were approximately 300 billion dollars greater than private consumer payments.   True   False

6. According to Graph A, the smallest increase in third party payments occurred from

    a. 1980 to 1990   **c.** 1992 to 1994
    b. 1990 to 1992   **d.** 1994 to 1996

7. What was the approximate total health care expenditures paid by both third party and private consumer sources in 1992?

    a. $700 billion   **d.** $300 billion
    b. $750 billion   **e.** $900 billion
    c. $850 billion

8. From 1994 to 1996, total expenditures by third parties and private consumers rose by approximately what amount?

    a. $10 billion   **c.** $100 billion
    b. $50 billion   **d.** $150 billion

9. According to Graph A, private consumers paid how many more dollars for health care in 1996 than in 1980?

    a. $70 billion   **d.** $50 billion
    b. $90 billion   **e.** $200 billion
    c. $110 billion

10. Which statement best describes Graph A?

    a. Health care expenses are at an all time low.
    b. Private consumers should pay more of the cost of health care.
    c. Third party payments rose steadily between 1990 and 1996, while private consumer payments remained the same.
    d. Expenditures by third parties and private consumers rose steadily between 1990 and 1996.
    e. Expenditures by private consumers have consistently been greater than expenditures by third party sources.

GRAPH B

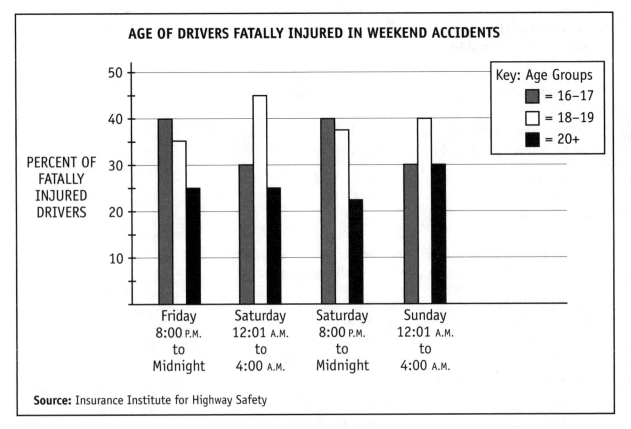

AGE OF DRIVERS FATALLY INJURED IN WEEKEND ACCIDENTS

Key: Age Groups
= 16–17
= 18–19
= 20+

PERCENT OF FATALLY INJURED DRIVERS

Friday 8:00 P.M. to Midnight

Saturday 12:01 A.M. to 4:00 A.M.

Saturday 8:00 P.M. to Midnight

Sunday 12:01 A.M. to 4:00 A.M.

Source: Insurance Institute for Highway Safety

Answer each question by filling in the blank, answering true or false, or choosing the best multiple-choice response.

1. Graph B represents drivers who are _____ in weekend car accidents.

2. Teenagers 16 and 17 years of age account for _____ percent of driver deaths on Saturday nights between 8:00 P.M. and midnight.

3. The same percent of 16- and 17-year-old drivers are fatally injured from 8:00 P.M. to midnight Friday and from 8:00 P.M. to midnight Saturday.   True   False

4. After midnight Saturday, more 16- and 17-year-old drivers have accidents than do drivers in the other two age groups.   True   False

5. 16- and 17-year-old drivers have the same percentage of fatal accidents as 20+ year-old drivers on

    a. Friday 8 P.M. to midnight
    b. Saturday 12:01 A.M. to 4:00 A.M.
    c. Saturday 8 P.M. to midnight
    d. Sunday 12:01 A.M. to 4:00 A.M.

6. The highest percentage of fatally injured drivers, 20 years and older, is _____ and occurs on Sunday 12:01 A.M. to 4:00 A.M.

    a. 55%
    b. 30%
    c. 15%
    d. 25%
    e. 32%

7. The time period when the highest percentage of fatally injured drivers is in the 18- to 19-year-old driving group is

    a. Friday 8 P.M. to midnight
    b. Saturday 12:01 A.M. to 4:00 A.M.
    c. Saturday 8 P.M. to midnight
    d. Sunday 12:01 A.M. to 4:00 A.M.

8. Of the groups listed, which has an average accident rate of 35% for the entire weekend?

    a. 16–17 year olds
    b. 18–19 year olds
    c. 20+ year olds

9. Which statement best describes Graph B?

    a. Fatally injured drivers are most often 16 to 17 years old for every time period during the weekend.
    b. Teenagers always have more fatal accidents than adults.
    c. On Friday night between 8:00 P.M. and midnight, 85% of all drivers who are fatally injured are 18 to 19 years old.
    d. Of the three age groups, drivers who are 20 years old and older are the least likely to be fatally injured in a weekend car accident.

Line Graphs

Line graphs are especially useful in showing trends and developments over a period of time. A line graph can be drawn with either curved or straight lines that extend across the graph in a horizontal direction.

Graph A shows how the number of people living at or below poverty levels changed over a period of several years.

GRAPH A

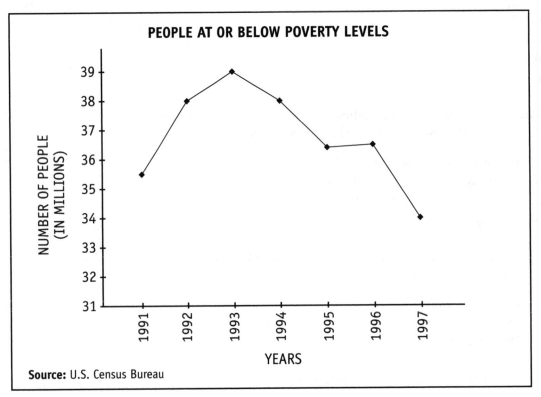

To answer questions about a line graph, follow the steps below.

Scanning the Graph

To scan a line graph, find the graph title, the axes names, and the labeled points along each axis.

EXAMPLE   Graph A shows the number of people living below poverty level from 1991 to _____.

STEP 1   Find the horizontal axis title: Years.

STEP 2   Find the last year included on the graph.

ANSWER: 1997

Reading the Graph

To read a line graph, find the information line, and read the labeled points along the horizontal and vertical axes.

EXAMPLE   In 1997, 30 million people were living at or below poverty   True   False
level.

> **STEP 1**   Find 1997 on the horizontal axis. From the bottom of the graph, move directly upward to the information line.

> **STEP 2**   From this point on the information line, move to the left to the labeled point on the vertical axis. You might use the edge of a paper to make the alignment accurately.

> **STEP 3**   Read the labeled point on the vertical axis—34.

ANSWER: False. In 1997, 34 million (not 30 million) people were living below poverty level.

Comprehension Questions

Inferences and predictions can be made by comparing values represented on the information line.

EXAMPLE   According to Graph A, the number of people living in poverty was the highest during which three years?

> **a.** 1991, 1992, 1993
> **b.** 1992, 1993, 1994
> **c.** 1993, 1994, 1995
> **d.** 1994, 1995, 1996
> **e.** 1995, 1996, 1997

> **STEP 1**   Scan across the information line from left to right. Identify the three consecutive points on the information line that are higher than any others.

> **STEP 2**   For each of these three points, move directly downward to the horizontal axis.

> **STEP 3**   Read the years labeled directly below the designated points on the information line.

ANSWER: b. 1992, 1993, 1994

Practice Graph I

Practice Graph I uses a single line to show the rise and fall of milk prices.

GRAPH I

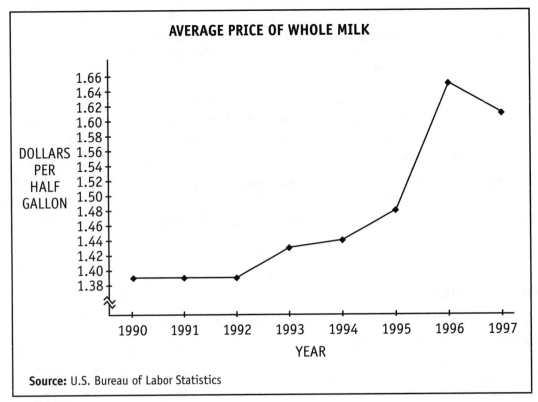

Scanning Graph I

Fill in each blank as indicated.

1. Graph I shows the average price of _____ _____.

2. The graph measures the price of milk in _____ per _____.

3. The graph shows the rise and fall of whole milk prices from _____ to _____.

Reading Graph I

Decide whether each statement is true or false and circle your answer.

4. In 1993, a half gallon of milk would have cost $1.52.   True   False

5. Between 1993 and 1996, whole milk prices rose.   True   False

6. The decrease in prices took place between 1996 and 1997.   True   False

Comprehension Questions

Answer each question by choosing the best multiple-choice response.

7. What is the difference in price between a half gallon of milk in 1994 and in 1995?

    a. $0.04
    b. $0.10
    c. $0.13
    d. $0.29
    e. $0.30

8. Between the years 1990 to 1997, what is the difference between the highest and the lowest whole milk prices shown on the graph?

    a. $0.22
    b. $0.15
    c. $0.26
    d. $0.06
    e. $0.03

9. From 1994 to 1997, milk prices

    a. rose steadily
    b. fell steadily
    c. rose sharply, then fell
    d. fell sharply, rose, then fell sharply again
    e. fell sharply, then rose steadily

10. From the information on the graph, you could infer the following:

    a. Farmers went on strike in 1991.
    b. The price of a half gallon of whole milk made its greatest price increases during the mid 1990's.
    c. The price of milk will rise throughout the 2000's.
    d. Milk prices reflect the rate of customer satisfaction.
    e. The fluctuation in milk prices has been a major cause of inflation and recession.

Practice Graph II

To compare different types of information, a **double line graph** uses more than one line. To prevent confusion, the lines are often drawn differently. A key is used to indicate the meanings of the different lines.

On some line graphs, the vertical axis line may appear to be broken and some values skipped. This is done to omit values that are not important and to save space. On Graph II, the values between 0 and 1.08 have been omitted.

GRAPH II

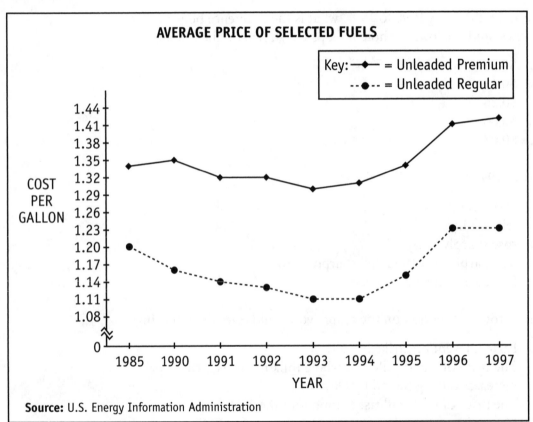

Source: U.S. Energy Information Administration

Scanning Graph II

Fill in each blank as indicated.

1. On Graph II, the solid line represents _____ fuel.

2. The average prices of fuels are measured in _____ per _____.

3. Graph II shows gasoline prices for _____ years.

Reading Graph II

Decide whether each statement is true or false and circle your answer.

4. In 1991 unleaded regular fuel cost an average of $1.32 per gallon.   True   False

5. The price of unleaded regular fell $0.20 per gallon from 1985   True   False
   to 1993.

6. In 1996, 10 gallons of unleaded premium would have cost $14.10.   True   False

Comprehension Questions

Answer each question by choosing the best multiple-choice response.

7. According to the graph, the highest priced fuel has consistently
   been

   a. leaded regular
   b. unleaded regular
   c. unleaded premium
   d. imported
   e. premium

8. Between 1990 and 1993, the prices of both fuels

   a. stayed the same
   b. rose steadily
   c. doubled
   d. decreased
   e. were cut by half

9. Between _____, the price of unleaded regular increased
   dramatically.

   a. 1994 and 1997
   b. 1992 and 1995
   c. 1985 and 1992
   d. 1994 and 1996
   e. 1993 and 1995

10. The following statement is true according to Graph II:

    a. Fuel prices will continue to rise throughout the 1990's.
    b. Fuel prices will fall during the 1990's.
    c. Fuel prices fluctuate depending upon supply and demand.
    d. Fuel prices fall most drastically during election years.
    e. Fuel prices have shown significant increases toward the end of
       the 1990's.

Practice Graph III

Graph III shows the consumer price indexes (CPI) for three categories of living expenses. Consumer price indexes are used by government and business to show the relative value of items in today's economy. The consumer price index below is based on what $100 would buy in 1982–1984. For example, if clothing has a CPI of 127 today, it would cost $127 today to buy goods that cost an average of $100 in 1982–1984.

GRAPH III

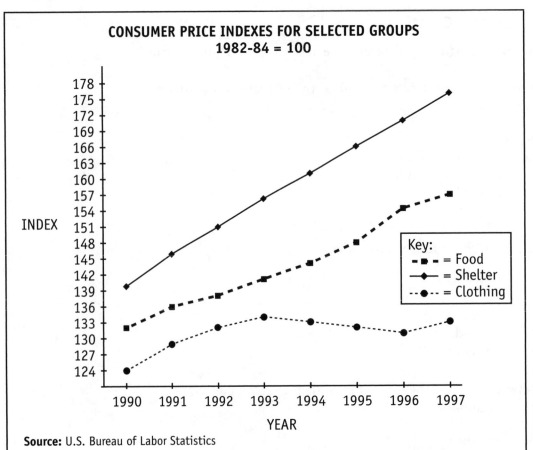

Source: U.S. Bureau of Labor Statistics

Scanning Graph III

Fill in each blank as indicated.

1. Graph III represents the consumer price indexes for
    _____, _____, and _____.

2. The consumer price index used in Graph III is based on costs in the
    years _____.

3. According to the key, --●-- stands for _____.

Reading Graph III

Decide whether each statement is true or false and circle your answer.

4. Shelter has a CPI of 148 for 1995.   True   False

5. It would have cost $136 in 1991 to buy the same food products   True   False
   that cost $100 in 1982–1984.

6. The 1990 consumer price index for food and shelter was   True   False
   approximately the same.

Comprehension Questions

Answer each question by choosing the best multiple-choice response.

7. For the years 1990 through 1997, the CPI that increased the most
   according to Graph III was in what area?

   a. shelter   **d.** fuel
   b. food   **e.** insurance
   c. clothing

8. What was the total increase in the consumer price index for food
   from 1991 to 1997?

   a. 13   **d.** 54
   b. 100   **e.** 21
   c. 37

9. In what year did the CPI for shelter begin to rise more rapidly than
   the consumer price indexes for food and clothing?

   a. 1991   **d.** 1994
   b. 1992   **e.** 1995
   c. 1993

10. The following statement is true according to Graph III:

    a. During the 1990's, the cost of shelter, food, and clothing were
       increasing at about the same rate.
    b. During the 2000's, clothing costs will always be cheaper than
       housing and food.
    c. By comparison, it takes an increased amount of dollars in the
       1990's to buy the same products purchased in the 1980's.
    d. The consumer price indexes for food, shelter, and clothing will
       decrease because of recession.
    e. Consumer price indexes rise most significantly during years
       of inflation.

Line Graphs: Applying Your Skills

The cost of raising a child has risen greatly during the last several years, and this trend is expected to continue from 2000 onward.

> **COSTS OF RAISING CHILDREN ON THE INCREASE**
> In the new millennium, costs for health care, food, clothing, shelter, and education are expected to continue to increase. Experts have projected that by the year 2014, it will cost between $200,000 and $300,000 to raise a child to age 17.

GRAPH A

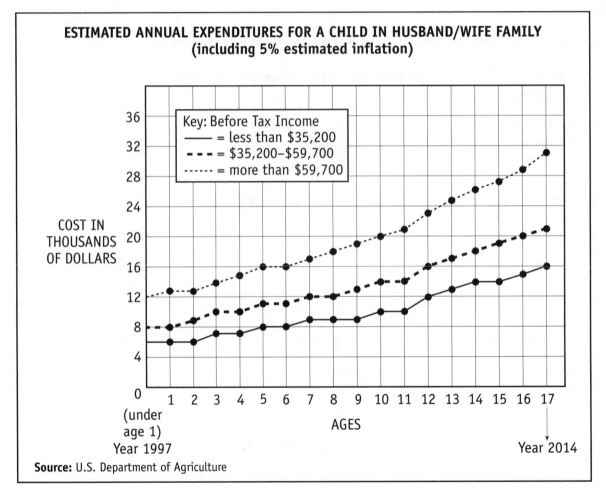

ESTIMATED ANNUAL EXPENDITURES FOR A CHILD IN HUSBAND/WIFE FAMILY
(including 5% estimated inflation)

Source: U.S. Department of Agriculture

Answer each question by filling in the blank, answering true or false, or choosing the best multiple-choice response.

1. Graph A represents the cost of raising a child to age _____.

2. The projected child-raising costs shown on Graph A include an estimated inflation rate of _____.

3. Overall, the yearly cost of raising a child is expected to increase as   True   False
   the child gets older.

4. The cost of raising a child from the ages of 11 to 17 in a family   True   False
   with before tax income of $48,000 is more than what it costs to
   raise a child from birth to age 10 in that same income bracket.

5. Graph A shows that the yearly cost of raising a child at age 14 is   True   False
   $11,000.

6. The period during which the cost of raising a child increases most
   rapidly across all three income groups is from the ages of

   a. 0 to 9
   b. 9 to 17

7. If 10% of the yearly cost of raising a child is for clothing, the cost of
   clothing a child at age 13 in a family with less than $35,200 income
   will be

   a. $1,200   d. $2,400
   b. $1,300   e. $1,400
   c. $1,700

8. According to Graph A, the cost of raising a child at age 13 is
   approximately _____ more for a family with income of $70,000
   than for a family with income of $55,000.

   a. $15,000   d. $4,000
   b. $16,000   e. $20,000
   c. $10,000

9. If the total cost of raising a child to age 17 is approximately
   $242,890, the estimated cost of 4 additional years in college, based
   on the formula of (cost of 1st 17 years) + ($20,000/year/college)
   would be estimated at

   a. $342,890   d. $322,890
   b. $312,890   e. $262,890
   c. $282,890

10. All the following statements can be concluded from Graph A EXCEPT:

   a. By the time a child is 17 years old, it is expected that he or she
       could cost at least $170,000 to feed, clothe, and educate.
   b. As a child gets older, the cost of raising the child gets higher.
   c. From 0 to 2, the cost of raising a child is relatively constant.
   d. The cost of raising a child in the first 5 years does not increase
       as rapidly as it does from 13 to 17.
   e. Inflation accounts for some of the increased costs of raising a child.

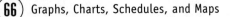

> **SAVE MONEY: WEATHERIZE YOUR HOME**
> Personal incomes are not increasing at the same rate as utility bills. Therefore, many families are cutting costs by taking steps to insulate and weatherize. The possible savings in weatherizing a home can make the money spent worthwhile.

GRAPH B

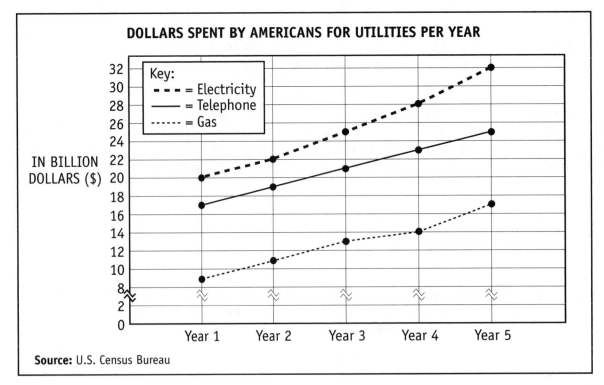

Source: U.S. Census Bureau

Answer each question by filling in the blank, answering true or false, or choosing the best multiple-choice response.

1. Graph B compares the money spent by Americans
   for _____, _____, and _____.

2. According to Graph B, Americans spend more money
   on _____ than on gas or _____.

3. In Year 1 the total money spent for the three utilities was   True   False
   approximately $32 billion.

4. In Year 2, Americans spent approximately $22 billion for   True   False
   telephone services.

5. From Graph B, you could conclude that heating costs are rising.   True   False

6. The greatest increase in dollars spent for electricity was from

   a. Year 1 to Year 2   d. Year 4 to Year 5
   b. Year 2 to Year 3   e. Year 5 to Year 6
   c. Year 3 to Year 4

7. In Year 3, _____ more was spent for electricity than for gas.

   a. $10 billion   d. $16 billion
   b. $12 billion   e. $18 billion
   c. $14 billion

8. In Year 4 the total dollar amount spent for electricity was two times
   the amount spent for

   a. gas   d. all utilities
   b. telephone   e. none of the above
   c. both gas and telephone

9. Which diagram most accurately describes Graph B?

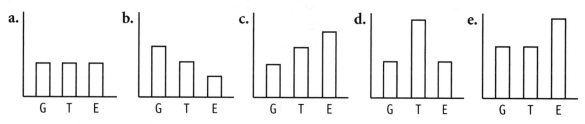

10. Which conclusion can be drawn from Graph B?

    a. American consumers are using more utilities every year.
    b. American consumers are spending a larger part of their income
       for utilities every year.
    c. Prices for utilities have increased more than prices in any other
       consumer area.
    d. The cost of personal telephone bills has kept up with the rate
       of inflation.
    e. Less money is spent by American consumers for gas than for
       other utilities.

Graph Review

Do all the following problems. Work accurately, but do not use outside help. After completing the review, check your answers with the key at the back of the book.

GRAPH A

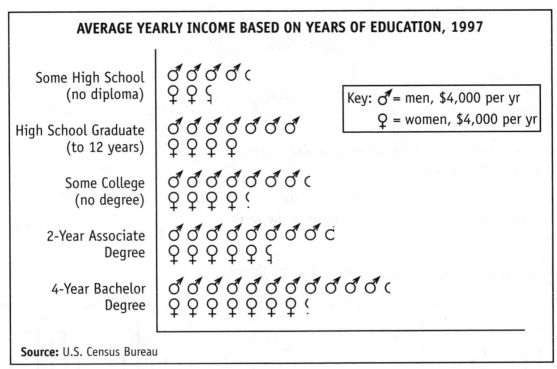

AVERAGE YEARLY INCOME BASED ON YEARS OF EDUCATION, 1997

Some High School (no diploma)

High School Graduate (to 12 years)

Some College (no degree)

2-Year Associate Degree

4-Year Bachelor Degree

Key: ♂ = men, $4,000 per yr
♀ = women, $4,000 per yr

Source: U.S. Census Bureau

Answer each question by filling in the blank, answering true or false, or choosing the best multiple-choice response.

1. Graph A shows income earned based on the amount of _____ a person has received.

2. From Graph A, you could infer that education helps to increase a person's chance for increased _____.

3. Men with a college bachelor degree make an average salary of over $40,000 per year.　　　　　　True　　　False

4. Women with some high school education earn an average of $8,000 per year.　　　　　　True　　　False

5. A female high school graduate earns about _____ than a male high school graduate.

   a. $12,000 less
   b. $9,000 more
   c. $12,000 more
   d. $6,000 less
   e. the same amount

6. On the average, the monthly income of a male high school graduate is about _____ more than a male who has some high school education.

   a. $4,000
   b. $10,000
   c. $20,000
   d. $8,000
   e. $12,000

7. On the average, a woman with a 4-year college degree will earn _____ more per year than a man with a high school diploma.

   a. $200
   b. $5,000
   c. $1,000
   d. $8,000
   e. $10,000

8. Which statement best describes Graph A?

   a. The amount of education individuals receive has little effect on their income opportunities.
   b. The amount of education individuals receive tends to improve their income opportunities.
   c. Even with an equivalent educational level, women tend to earn more money than men.

GRAPH B

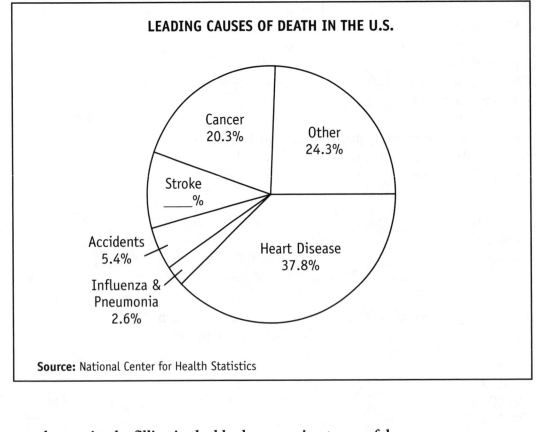

LEADING CAUSES OF DEATH IN THE U.S.

Cancer 20.3%

Other 24.3%

Stroke ____%

Accidents 5.4%

Influenza & Pneumonia 2.6%

Heart Disease 37.8%

Source: National Center for Health Statistics

Answer each question by filling in the blank, answering true or false, or choosing the best multiple-choice response.

9. Graph B represents the leading causes of _____ in the _____.

10. The percentage of deaths in the United States due to strokes is _____. (Fill this in on the graph.)

11. Heart disease and accidents are the two leading causes of death in the United States.   True   False

12. Each year heart disease accounts for slightly more than one-third of the deaths in the United States.   True   False

13. The number of deaths due to accidents is _____ less than deaths due to strokes.

   a. 4.2%
   b. 3.6%
   c. 1.8%
   d. 17.8%
   e. 13.5%

14. Together, heart disease and _____ account yearly for slightly more than one-half of the deaths in the United States.

    a. heart attacks
    b. strokes
    c. cancer
    d. pneumonia
    e. other

15. A diagram that correctly shows the relationship between two causes of death is:

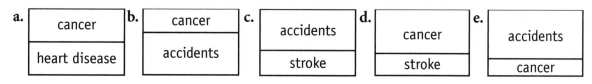

16. The following statements can be concluded from Graph B EXCEPT:

    a. Cancer and heart disease cause more than 50% of the deaths in the United States.
    b. Fewer people die from accidents each year in the United States than from strokes.
    c. The leading cause of death in the United States is heart disease.
    d. Graph B shows five leading causes of death in the United States.
    e. Research funding for heart disease exceeds the funding for all other fatal illnesses.

GRAPH C

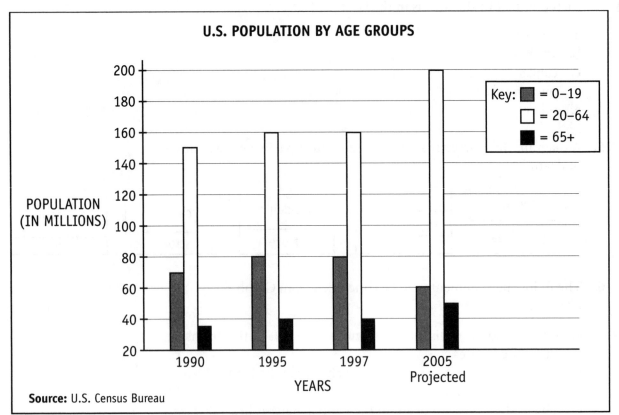

U.S. POPULATION BY AGE GROUPS

Key: ■ = 0–19
□ = 20–64
■ = 65+

POPULATION (IN MILLIONS)

YEARS

2005 Projected

Source: U.S. Census Bureau

Answer each question by filling in the blank, answering true or false, or choosing the best multiple-choice response.

17. Graph C represents the population of the _____, and for each year shown, the population is divided into _____ age groups.

18. In 1990 the total population in the United States was _____.

19. Graph C shows the world population over a 22-year period.   True   False

20. By the year 2005, it is estimated that there will be about 200 million people who are 20 to 64 years old.   True   False

21. Between 1990 and 1995, approximately what was the total increase in population in the United States?

   a. 45 million
   b. 2 million
   c. 15 million
   d. 20 million
   e. 65 million

22. Between 1990 and 2005, it is estimated that the _____ age group will show the greatest gain in total number of people.

    a. 0–19
    b. 20–64
    c. 65+

23. Between 1990 and 2005, it is estimated that the number of people in the 65+ age group will

    a. increase rapidly
    b. decrease rapidly
    c. consistently increase
    d. increase until 1997

24. Which conclusion can be drawn from Graph C?

    a. The life expectancy of Americans is increasing.
    b. Since 1995 older people are retiring at an earlier age than before 1990.
    c. The total number of people in each age group shown is expected to be greater in the year 2005 than in any previous year.
    d. There were more teenagers in 1990 than the projections show for 2005.

GRAPH D

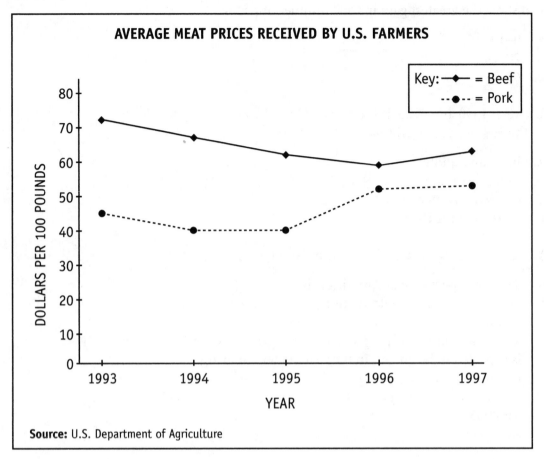

Source: U.S. Department of Agriculture

Answer each question by filling in the blank, answering true or false, or choosing the best multiple-choice response.

25. Graph D shows the average prices received by U.S. farmers for
    _____ and _____.

26. The highest price paid for beef occurred in the year _____
    and was equal to $_____ per 100 pounds.

27. The prices appearing in Graph D represent the number of   True   False
    dollars for every 10 pounds of meat.

28. Prices for pork rose steadily from the years 1993 to 1997.   True   False

29. During what year did the farmers receive $52 per 100 pounds of pork?

   a. 1997
   b. 1993
   c. 1994
   d. 1995
   e. 1996

30. During 1995, a farmer could earn approximately _____ more for 100 pounds of beef than for 100 pounds of pork.

   a. $20
   b. $30
   c. $5
   d. $10
   e. $15

31. Which diagram best represents information from Graph D?

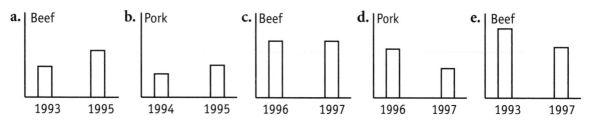

32. Which conclusion can be drawn from Graph D?

   a. Meat prices declined during the 1990's.
   b. In the 1990's, farmers earned more income from beef than from pork.
   c. Pork and beef prices will rise and fall with the consumer price index.
   d. The price that farmers receive for pork can be expected to rise above the price of beef during the next several years.
   e. The prices a consumer pays for beef and pork decrease as the average income received by U.S. farmers decreases.

Graph Review Chart

Circle the number of any problem that you missed and review the appropriate pages. A passing score is 27 correct answers. If you miss more than 5 questions, you should review this chapter.

PROBLEM NUMBERS	SKILL AREA	PRACTICE PAGES
1, 2, 3, 4, 5, 6, 7, 8	pictograph	16–31
9, 10, 11, 12, 13, 14, 15, 16	circle graph	16–19, 32–43
17, 18, 19, 20, 21, 22, 23, 24	bar graph	16–19, 44–55
25, 26, 27, 28, 29, 30, 31, 32	line graph	16–19, 56–67

SCHEDULES AND CHARTS

Schedule and Chart Skills Inventory

This inventory allows you to measure your skills in reading and interpreting schedules and charts. Correct answers are listed by page number at the back of the book.

CHART I

Time Difference From U.S. Mainland to:	U.S. Time Zones in Standard Time			
	Eastern (New York)	Central (Chicago)	Mountain (Denver)	Pacific (Los Angeles)
Paris (France)	6	7	8	9
Munich (Germany)	6	7	8	9
Athens (Greece)	7	8	9	10
Peking (China)	13	14	15	16
Tel Aviv (Israel)	7	8	9	10
Tokyo (Japan)	14	15	16	17
Auckland (New Zealand)	18	19	20	21
Durban (South Africa)	7	8	9	10
Stockholm (Sweden)	6	7	8	9
London (England)	5	6	7	8
Caracas (Venezuela)	1	2	3	4

International Time Table: Time Difference in Phone Dialing**

**To compute time change, add the number of hours shown to your watch.

Answer each question below.

1. There are four United States time zones listed on Chart I: Eastern, _____, _____, and Pacific.

2. If you drive from the Eastern time zone to the Pacific time zone, you would need to change your watch by _____ hours.

3. The time difference between Chicago and Tokyo is 19 hours.   True   False

4. When it is 3 o'clock P.M. in New York, it is 8 o'clock P.M. in London.   True   False

5. When it is 12 o'clock noon Tuesday in Denver, it is 8 o'clock _____ in Auckland, New Zealand.

   a. Tuesday morning   c. Wednesday evening
   b. Tuesday evening   d. Wednesday morning

SCHEDULE I

Parcel Post Rate Schedule
[for packages over 1 pound]

Weight Not Over In Pounds (lb)	ZONES BASED ON MILES PARCEL IS SHIPPED							
	Miles Zone: Local	up to 150 1 & 2	151–300 3	301–600 4	601–1000 5	1001–1400 6	1401–1800 7	1801+ 8
1.5	$1.14	$1.54	$1.57	$1.63	$1.72	$1.81	$1.92	$2.02
2.0	1.16	1.57	1.61	1.69	1.81	1.93	2.08	2.21
2.5	1.18	1.60	1.66	1.76	1.90	2.06	2.24	2.40
3.0	1.20	1.63	1.70	1.82	1.99	2.18	2.40	2.60
3.5	1.22	1.66	1.74	1.88	2.08	2.30	2.56	2.79
4.0	1.24	1.70	1.79	1.94	2.18	2.42	2.72	2.98
4.5	1.26	1.73	1.83	2.01	2.27	2.55	2.88	3.17
5.0	1.28	1.76	1.88	2.07	2.36	2.67	3.05	3.37
6.0	1.31	1.82	1.96	2.20	2.54	2.92	3.37	3.75
7.0	1.35	1.89	2.05	2.32	2.73	3.16	3.69	4.14
8.0	1.39	1.95	2.14	2.45	2.91	3.41	4.01	4.52
9.0	1.43	2.02	2.22	2.57	3.10	3.65	4.33	4.91
10.0	1.47	2.08	2.31	2.70	3.28	3.90	4.65	5.29
11.0	1.51	2.14	2.40	2.83	3.46	4.15	4.97	5.68
12.0	1.55	2.21	2.48	2.95	3.65	4.39	5.29	6.06
13.0	1.59	2.27	2.57	3.08	3.83	4.64	5.61	6.45
14.0	1.63	2.34	2.66	3.20	4.02	4.88	5.93	6.83
15.0	1.67	2.40	2.75	3.33	4.20	5.13	6.26	7.22

Source: United Parcel Service

Answer each question by filling in the blank, answering true or false, or choosing the best multiple-choice response.

6. Parcel post rates are determined by the _____ in which the parcel is shipped and the _____ of the parcel.

7. Schedule I gives the cost of mailing a package that weighs not more than _____ pounds.
   (number)

8. According to Schedule I, mailing a 7-pound 4-ounce package to a city 850 miles away costs $2.91.   True   False

9. The postage for mailing a package to a city 350 miles away is listed under Zone 5.   True   False

10. The postage to mail a 10-pound package 510 miles is
_____ more than a 12-pound package 225 miles away.

   a. 99¢
   b. 22¢
   c. 54¢
   d. 11¢
   e. 89¢

11. The cost of mailing two 5-pound packages 525 miles is _____
more than mailing one 10-pound package the same distance.

   a. $1.64
   b. $1.57
   c. $2.20
   d. $1.44
   e. $3.25

Schedule and Chart Skills Inventory Chart

Use this inventory to see what you already know about schedules and charts and what you need to work on. A passing score is 9 correct answers. Even if you have a passing score, circle the number of any problem that you miss and turn to the practice pages indicated for further instruction.

PROBLEM NUMBERS	SKILL AREA	PRACTICE PAGES
1, 2, 3, 4, 5	chart	80–87, 92–95
6, 7, 8, 9, 10, 11	schedule	80–85, 88–95

What Are Schedules and Charts?

Schedules and charts show specific values by listing numbers and words in columns and rows.

Comparing Schedules and Charts

Look at the examples of a schedule and a chart below. Schedules and charts may look very much alike. However, they may have different uses.

EXAMPLE   Bus Schedule

Bus Schedule ♿
#22 LCC EXPRESS

LEAVE 10th & Willamette	19th & Pearl	30th & Hilyard	ARRIVE Lane Commnty College
7:05	7:09	7:13	7:21
7:35	7:39	7:43	7:51
8:05	8:09	8:13	8:21
8:35	8:39	8:43	8:51
9:05	9:09	9:13	9:21
9:35	9:39	9:43	9:51
10:05	10:09	10:13	10:21
10:35	10:39	10:43	10:51
11:05	11:09	11:13	11:21
11:35	11:39	11:43	11:51
12:05	12:09	12:13	12:21
12:35	12:39	12:43	12:51
1:05	1:09	1:13	1:21
1:35	1:39	1:43	1:51
2:05	2:09	2:13	2:21
2:35	2:39	2:43	2:51
3:05	3:09	3:13	3:21
3:35	3:39	3:43	3:51
4:05	4:09	4:13	4:21
4:35	4:39	4:43	4:51

Schedules often list times of events. Therefore, a useful definition of a schedule is

> a list of facts and relations primarily dealing with *times* of events.

Examples of other schedules:

    train
    television
    sports events
    school registration

EXAMPLE   Nutrition Chart

NUTRITION CHART OF FOOD GROUPS FOR 3–OUNCE SERVINGS

Food	Calories	Protein (g)
Chicken	116	20
Lamb	348	17
Beef	165	25
Ham	245	18
Pork	339	21

Source: U.S. Department of Agriculture

Charts are used to compare values of one item with another. Therefore, a useful definition of a chart is

> a list of information on which *items* are compared.

Examples of other charts:

    calorie
    weight
    mileage
    medical costs

Parts of Schedules and Charts

Schedules and Charts Have Titles

The **title** is a short description of the topic or main idea.

EXAMPLE   Bus Schedule

> ## Bus Schedule ♿
> #22 LCC EXPRESS

The title gives the bus route number (22) and the bus route name (LCC Express).

EXAMPLE   Nutrition Chart

> **NUTRITION CHART OF FOOD GROUPS FOR 3-OUNCE SERVINGS**

The title gives the topic: nutrition of food groups.

Schedules and Charts Often Contain Tables

A **table** is a list of words and numbers written in rows and columns. Columns are read up and down. Rows are read across. (You will learn how to read tables on pages 84 and 85.)

LEAVE 10th & Willamette	19th & Pearl	30th & Hilyard	ARRIVE Lane Commnty College
7:05	7:09	7:13	7:21
7:35	7:39	7:43	7:51
8:05	8:09	8:13	8:21
8:35	8:39	8:43	8:51
9:05	9:09	9:13	9:21
9:35	9:39	9:43	9:51
10:05	10:09	10:13	10:21
10:35	10:39	10:43	10:51
11:05	11:09	11:13	11:21
11:35	11:39	11:43	11:51
12:05	12:09	12:13	12:21
12:35	12:39	12:43	12:51
1:05	1:09	1:13	1:21
1:35	1:39	1:43	1:51
2:05	2:09	2:13	2:21
2:35	2:39	2:43	2:51
3:05	3:09	3:13	3:21
3:35	3:39	3:43	3:51
4:05	4:09	4:13	4:21
4:35	4:39	4:43	4:51

Food	Calories	Protein (g)
Chicken	116	20
Lamb	348	17
Beef	165	25
Ham	245	18
Pork	339	21
Source: U.S. Department of Agriculture		

This table compares nutritional values of five food groups.

This table gives the times when the bus leaves and arrives at specific locations.

Schedules and Charts Often Have Symbols

Symbols provide additional information. A **key** is often used to give the meaning or value of a symbol.

♿   This symbol means that the bus has a lift to pick up riders in wheelchairs.

(g)   This symbol stands for the word *grams*, the weight unit that is used to measure protein.

Types of Questions

In this section of the workbook, you will be asked three types of questions to find information on charts and schedules.

Scanning the Chart (Schedule) Questions

"Scanning the Chart (Schedule) Questions" requires looking at the main topics and information on a chart or schedule. These questions are answered by filling in words to complete the sentence. In scanning a chart (schedule),

- Pay attention to the titles, names of columns and rows, and identified units of measurement.
- Check the source of the chart's (schedule's) information.

EXAMPLE   Chart A gives the recommended daily dietary allowance of Vitamin _____.

ANSWER: C. Scan the titles of the columns for this information.

Reading the Chart (Schedule) Questions

"Reading the Chart (Schedule) Questions" involve locating specific information on the chart or schedule. When reading a chart (schedule),

- Read horizontally (from left to right) and vertically (from top to bottom).

EXAMPLE   It is recommended that a boy, age 12, consume 56 grams of protein daily.
True   False

ANSWER: False. Starting from the left side of the chart and at the row title Males 11–14, read across to 45 in the protein column.

Comprehension Questions

Comprehension questions require you to compare values, make inferences, and draw conclusions.

EXAMPLE   Women 15–18 years old require _____ milligrams of Vitamin C than women 11–14 years old.

a. 10 more   c. 50 more
b. 10 less   d. 50 less

ANSWER: a. 10 more. Subtract the amount of Vitamin C for women 11–14 (50 mg) from the number for women 15–18 (60 mg).

CHART A

RECOMMENDED DAILY DIETARY ALLOWANCES

	Age (years)	Weight (lb)	Height (in.)	Protein (g)	Vitamin C (mg)
Infants	0.0–0.5	13	24	kg x 2.2	30
	0.5–1.0	20	28	kg x 2.0	35
Children	1–3	29	35	16	40
	4–6	44	44	24	45
	7–10	62	52	28	45
Males	11–14	99	62	45	50
	15–18	145	69	59	60
	19–22	154	70	58	60
	23–50	154	70	63	60
	51+	154	70	63	60
Females	11–14	101	62	46	50
	15–18	120	64	44	60
	19–22	120	64	46	60
	23–50	120	64	50	60
	51+	120	64	50	60
Pregnant				60	70
Lactating				65	90

Source: National Academy of Sciences, 1995

Use Care in Reading Questions

Careful reading may help you to answer "trickier" questions. The hints below will alert you to possible problem areas.

Information Is True but Not Contained on the Chart

EXAMPLE   According to Chart B, milk is more nutritional than candy.   True   False

ANSWER: False. While this fact is true, the chart does not compare the nutritional content of foods. Chart B only recommends the number of servings for certain food groups.

Words That Are Incorrectly Used as Possible Answers

EXAMPLE   Chart B tells
  a. the percentage of certain food groups recommended for different types of people
  b. the number of servings per food group recommended on a daily basis
  c. the percentage of servings per food group recommended on a daily basis

ANSWER: b. The chart is about numbers of servings. Nothing is said about percentages, so choices **a** and **c** can be eliminated.

Symbols, Equivalents, and Abbreviations Are Often Used to Get the Correct Answer

EXAMPLE   According to Chart B, the number of servings of meat per day on a Step I diet is _____ ounces.

ANSWER: Less than or equal to 6. In order to provide the correct answer, find the correct meaning of the symbol, ≤. This symbol is defined at the bottom of the chart. Second, careful reading requires identification of the ounces based on a Step I diet, not a Step II diet.

CHART B

Daily Food Guide

Food Group	Number of Servings	Serving Size
Lean meat, poultry, fish and shellfish	≤6 ounces a day on Step I diet ≤5 ounces a day on Step II diet (leanest cuts only)	
Skim/low fat dairy foods	2–3	1 cup skim or 1 percent milk 1 cup nonfat or low fat yogurt 1 ounce low fat or fat free cheese that has 3 grams of fat or less in a serving
Eggs	≤4 yolks a week on Step I diet* ≤2 yolks a week on Step II diet*	
Fats and oils	≤6–8*	1 teaspoon soft margarine or vegetable oil 1 tablespoon salad dressing 1 ounce nuts
Fruits	2–4	1 piece fruit ¹/₂ cup diced fruit ³/₄ cup fruit juice
Vegetables	3–5	1 cup leafy or raw ¹/₂ cup cooked ³/₄ cup juice
Breads, cereals, pasta, rice, dry peas and beans, grains, and potatoes	6–11	1 slice bread ¹/₂ bun, bagel, muffin 1 ounce dry cereal ¹/₂ cup cooked cereal, dry peas or beans, potatoes, or rice or other grains ¹/₂ cup tofu
Sweets and snacks	Now-and-then	

*Includes food preparation; for fats and oils also includes salad dressings and nuts.
≤ = less than or equal to
Source: American Heart Association, 1995

Schedules and Charts

Finding Information on a Table

To answer questions about schedules and charts, you will need to know how to read the columns and rows on the table. Look at the chart below.

NUTRITION CHART OF FOOD GROUPS FOR 3–OUNCE SERVINGS		
Food	Calories	Protein (g)
Chicken	116	20
Lamb	348	17
Beef	165	25
Ham	245	18
Pork	339	21

Source: U.S. Department of Agriculture

Follow the steps below to answer questions about charts and schedules.

EXAMPLE   How many calories are contained in a 3-ounce serving of ham?

STEP 1   The table lists both the calories and protein found in the five types of foods. The title states that these figures are given for 3-ounce servings.

Food	Calories	Protein (g)
Chicken	116	20
Lamb	348	17
Beef	165	25
Ham	245	18
Pork	339	21

STEP 2   Find the row labeled "Ham." Place a piece of paper under that row extending across the chart or use your finger as shown.

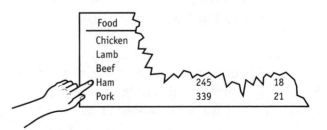

STEP 3 In the column headings, locate the title Calories. Move your eye (or finger) down that column until it meets the row labeled Ham.

Food	Calories	Protein (g)
Chicken	116	
Lamb	348	
Beef	165	
Ham	245	
Pork	339	

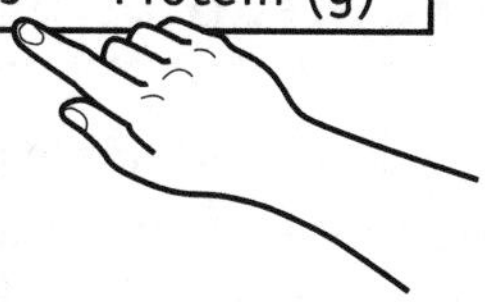

ANSWER: There are **245 calories** in a 3-ounce serving of ham.

Answer the questions below. Check your answers to see that you are reading the chart correctly. Remember to carefully follow the steps shown above.

EXAMPLE 1 There are seventeen grams of protein in a 3-ounce serving of
_____.

ANSWER: lamb. In this case, you locate the information within the table and find your answer as the row title. Find 17 in the protein column and then look across to the row labeled Lamb.

EXAMPLE 2 A 6-ounce serving of pork would contain 339 calories.　　　True　　　False

ANSWER: False. You must multiply the given number of calories in a 3-ounce serving (339) by 2 to get the total number of calories in a 6-ounce serving. (339 × 2 = 678)

Practice Chart I

Chart I is a table that contains information about the importance of serving size and amount of calories necessary to maintain a balanced diet and good health. Often charts have two or more sections from which information can be drawn.

CHART I

Serving Sizes

To make the most of maintaining a balanced diet, you need to know what counts as a serving.

Food Group	Serving Size
Bread, Cereal, Rice and Pasta	1 slice bread, 1 ounce ready-to-eat cereal, $\frac{1}{2}$ cup cooked cereal, rice or pasta, 5–6 small crackers
Vegetable	1 cup raw, leafy vegetables, $\frac{1}{2}$ cup cooked or chopped raw vegetables, $\frac{3}{4}$ cup vegetable juice
Fruit	1 medium apple, banana or orange, $\frac{1}{2}$ cup chopped, cooked, or canned fruit, $\frac{3}{4}$ cup fruit juice
Milk, Yogurt and Cheese	1 cup milk or yogurt, $1\frac{1}{2}$ ounces natural cheese, 2 ounces process cheese
Meat, Poultry, Fish, Dry Beans, Eggs and Nuts	1–3 ounces cooked lean meat, poultry or fish. Foods which count as 1 ounce of meat: $\frac{1}{2}$ cup cooked dry beans, 1 egg, 2 tablespoons peanut butter, $\frac{1}{3}$ cup nuts

Sample Food Plan Per Day: Calories

	1,600 For Many Sedentary Women and Some Older Adults	2,200 Most Children, Teenage Girls, Active Women and Many Sedentary Men	2,800 Teenage Boys, Many Active Men and Some Very Active Women
	Servings	Servings	Servings
Bread Group	6	9	11
Fruit Group	2	3	4
Vegetable Group	3	4	5
Milk Group	2–3	2–3	2–3
Meat Group	5 ounces	6 ounces	7 ounces

Source: U.S. Department of Health and Human Services

Scanning Chart I

Fill in each blank as indicated.

1. The food groups that are listed on Chart I are bread, fruit, _____, _____, and _____.

2. Chart I compares the daily recommended _____ for three groups of children and adults.

3. The top section of Chart I identifies the serving size for each of the five _____ groups.

Reading Chart I

Decide whether each sentence is true or false and circle your answer.

4. Active teenage girls require at least 1,600 calories per day.　　　　True　　　　False

5. An active teenage boy in high school is expected to consume about 1,200 calories more per day than his retired grandfather.　　　　True　　　　False

6. To consume 2,200 calories per day, one might consume at least 2 cups of milk, an 8-ounce steak, 1 apple, 1 cup of orange juice, 1 banana, a green salad for lunch and dinner (equivalent to 4 cups), 1 cup of cooked cereal, 4 slices of bread, and 20 crackers for snacks.　　　　True　　　　False

Comprehension Questions

Answer each question by choosing the best multiple-choice response.

7. Two servings of the milk group would consist of

    a. 3 slices of bread and 2 ounces of lean meat
    b. 1 cup of yogurt and 1 cup of milk
    c. 2 ounces of cheese and a carrot
    d. 1 cup of milk and $\frac{1}{2}$ cup of fruit

8. The amount of servings of bread for teenage boys is almost _____ the number of servings for sedentary women.

    a. same　　　　b. $\frac{1}{2}$　　　　c. two times　　　　d. three times

9. All persons are recommended to consume the same amount of food servings of the _____ group per day.

    a. meat　　　　b. milk　　　　c. fruit　　　　d. vegetable

10. From Chart I you can conclude the following:

    a. An increase in calories is needed as one gets older.
    b. Age and the activities of a person are the two best factors to determine the amount of calories needed per day.
    c. The more calories consumed per day, the more one must diet.

Practice Schedule I

Schedule I lists registration information for classes at a community college. On the basis of the first letter of his or her last name, a student reads down the column to find the appropriate row and then reads across the row to find the day, date, and time to sign up for classes.

SCHEDULE I

TO NEW AND RETURNING STUDENTS
REGISTRATION INFORMATION

Registration is Monday, Jan. 26 through Saturday, Jan. 31

PLEASE NOTE FOLLOWING SCHEDULE FOR OUR
STREAMLINED REGISTRATION PROCEDURES

If Your Last Name Begins with the Letter:	Please Register on:	During the Hours of:
S•T	Monday, January 26	9:00 A.M.•1:00 P.M.
C•D•E	Monday, January 26	1:00 P.M.•5:00 P.M.
O•P•Q•R	Tuesday, January 27	9:00 A.M.•1:00 P.M.
F•G•H	Tuesday, January 27	1:00 P.M.•5:00 P.M.
M•N	Wednesday, January 28	9:00 A.M.•1:00 P.M.
I•J•K•L	Wednesday, January 28	1:00 P.M.•5:00 P.M.
U•V•W•X•Y•Z	Thursday, January 29	9:00 A.M.•1:00 P.M.
A•B	Thursday, January 29	1:00 P.M.•5:00 P.M.

Open registration regardless of first letter of last name available daily between 5 and 6 P.M. and all day Friday, Jan. 30 and Saturday, Jan. 31 from 10 A.M. to 2 P.M.

JOHNSON COMMUNITY COLLEGE
927 WEBSTER ST.

FOR FURTHER INFORMATION PHONE

787–1984

Scanning Schedule I

Fill in each blank as indicated.

1. This schedule is useful for students enrolling in classes at
   _____ .

2. Student registrations begin on _____ and end on
   (day—date)
   _____ .
   (day—date)

3. To find out more information about registration, a student can call
   the phone number _____ .

Reading Schedule I

Decide whether each statement is true or false and circle your answer.

4. A new student, Henry Foster, can register on Tuesday, January 27,   True   False
    between 1:00 P.M. and 5:00 P.M.

5. Mercedes Rodriquez can register on Thursday before 5:00 P.M.   True   False

6. If Nate Fennerty does not register by Tuesday at 5:00 P.M., he will   True   False
    not be able to register during the semester.

Comprehension Questions

Answer each question by choosing the best multiple-choice response.

7. The total number of registration hours on Monday is

    a. 4
    b. 8
    c. 9
    d. 10
    e. none of the above

8. If a student misses his or her assigned registration time, the student
    can register

    a. between 5:00 and 6:30 P.M. daily
    b. on Saturday, February 1
    c. on Friday, January 30

9. Sue Smith will have _____ hours in which to enroll on January 26.

    a. 1
    b. 3
    c. 4
    d. 8
    e. 9

10. If 160 students enroll by 1:00 P.M. on Monday, the average number
    of registrations per hour would be

    a. 70
    b. 40
    c. 160
    d. 10
    e. 80

Practice Schedule II

Schedule II lists times that trains arrive at and leave cities between Sacramento and Los Angeles, California.

For ease of reading, origin and destination stations are indicated with a "Dp" for "departs" and an "Ar" for "arrives." In addition, some stations have an "Ar" and a "Dp" indicator to tell the reader how long the train is scheduled to remain in that station. At stations without an "Ar" or "Dp" indicator, the train stops only long enough to permit passengers to board safely.

Notice that small picture symbols tell the services available on this train. For example, ☕ means that this train has sandwich, snack, and beverage service.

SCHEDULE II

Western Schedules

Sacramento-Oakland-Los Angeles

Read Down				Read Up	
15		Train Number		**18**	
Daily		Frequency of Operation		**Daily**	
R ☕ 🛄 ⊠				R ☕ 🛄 ⊠	
7 55 P	Dp	**Sacramento, CA**	Ar	9 30 A	
8 13 P		Davis, CA		8 48 A	
8 41 P		Suisan-Fairfield, CA		8 21 A	
9 03 P		Martinez, CA		7 59 A	
9 32 P		Richmond, CA		7 28 A	
9 50 P	Ar	**Oakland, CA**	Dp	7 15 A	
10 10 P	Dp		Ar	7 00 A	
11 25 P	Ar	**San Jose, CA**	Dp	5 45 A	
11 30 P	Dp		Ar	5 41 A	
12 55 A		Salinas, CA		4 12 A	
3 51 A	Ar	San Luis Obispo, CA	Dp	1 28 A	
3 59 A	Dp	*(Hearst Castle)*	Ar	1 20 A	
6 17 A		Santa Barbara, CA		10 45 P	
7 06 A		Oxnard, CA		9 53 P	
8 18 A		Glendale, CA		8 43 P	
9 00 A	Ar	**Los Angeles, CA**	Dp	8 25 P	

Reference Marks

R All reserved train.

☕ Sleeping car service.

✓ Club car service.

⊠ Tray meal and beverage service.

☕ Sandwich, snack and beverage service.

🛄 Checked baggage handled.

A A.M.
P P.M.

Ar Arrive
Dp Depart

Scanning Schedule II

Fill in each blank as indicated.

1. The train stops at four cities between Sacramento and Oakland: Davis, _____, _____, and _____.

2. The number of the train going from Sacramento to Los Angeles is _____, and the number of the train going from Los Angeles to Sacramento is _____.

3. The symbol ✍ means that the train has _____.

Reading Schedule II

Decide whether each sentence is true or false and circle your answer.

4. Traveling from Sacramento, train #15 arrives in Los Angeles at 9:00 A.M. the next day.   True   False

5. A train to Los Angeles leaves San Jose at 7:00 A.M.   True   False

6. Train #18 has sleeping car service.   True   False

Comprehension Questions

Answer each question by choosing the best multiple-choice response.

7. The total time it takes to travel from Sacramento to Oakland is slightly less than

   a. 4 hours   **d.** 2 hours
   b. 9 hours   **e.** 1 hour
   c. 7 hours

8. If you leave Los Angeles on Friday evening at 8:25 P.M., you will arrive in Sacramento at 9:30

   a. Saturday evening   **d.** Saturday morning
   b. Sunday morning   **e.** Sunday evening
   c. Friday evening

9. Train #18 stops at Oakland for _____ minutes before it leaves for Sacramento.

   a. 30   **d.** 10
   b. 15   **e.** 12
   c. 20

10. It takes a little over _____ for train #15 to travel from Sacramento to Los Angeles.

   a. 25 hours   **d.** 8 hours
   b. 13 hours   **e.** 2 hours
   c. 5 hours

Schedules and Charts: Applying Your Skills

A single person must file a tax return, known as Form 1040, in each year that his or her income is at least $6,950 for the year. Likewise, a married person filing a return with his or her spouse must file if their income is at least $12,500.

Many people choose to have their taxes prepared by a professional. Chart I compares the times for preparation, quoted fees, and actual fees of various tax preparers. In addition, Chart I includes the total federal and state tax bill, which is the amount of tax each individual taxpayer, based on his income, pays to the state and federal governments.

CHART I

Comparing Tax Preparers					
Tax Preparation Firm or Agency	Quote Fee	Actual Fee	Total Federal and State Tax Bill	Time From Appointment to Receiving Return	Was It Longer Than They Estimated?
Hammer & Associates, Inc.	$40–45	$130	$6,822	Same day	Same day
H & R Block	38–40	96	6,160	19 days	9 days late
Sutton & Jones, Ltd., attorneys	200	200	6,462	16 days	6 days late
Zeus Insurance Agency	25–30	75	6,840	4 days	On time
Beneficial Tax Service	45–50	68	6,301	17 days	9 days late
Frank T. Madison, Certified Public Accountant	125	120	6,219	14 days	2 days early
Whitney & White, accountants	300–400	375	6,290	9 days	12 days early
Internal Revenue Service	0	0	7,155	Same day	Same day
Conner & Matson, accountants	250–400	350	6,206	23 days	6 days late
Jay A. Sullivan, attorney at law	25–50	100	6,673	7 days	5 days early

Answer each question below.

1. Chart I compares quoted fees with _____ fees charged by income tax preparers.

2. From Chart I, it can be concluded that many tax preparers are late in completing their clients' tax returns.   True   False

3. The IRS does not charge for tax preparation assistance.   True   False

4. For most firms listed, actual fees are less than quoted fees.   True   False

5. The two firms that provide same day service are

   a. Hammer and Zeus   **c.** Hammer and IRS
   b. IRS and Sullivan   **d.** Hammer and Sutton

6. According to Chart I, the difference between the quoted fee and the actual fee charged by H&R Block ranges from

   a. $29 to $49   **c.** $56 to $58
   b. $37 to $67   **d.** $38 to $68

7. If you had an appointment with the Zeus Insurance Agency on Friday, March 12th, you could expect your tax return by (**Note:** Days given on chart are working days.)

   a. Friday, March 12   **c.** Thursday, March 18
   b. Tuesday, March 16   **d.** Friday, March 19

8. Which diagram represents Chart I most accurately?

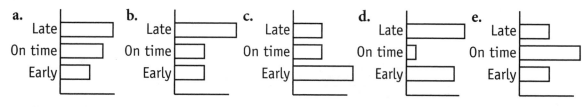

9. Which statement best describes Chart I?

   a. All tax preparation firms charge more than their quoted fees.
   b. The IRS does not charge for tax preparation assistance, and it calculates the lowest tax bill.
   c. Tax preparation firms vary widely in fees charged, total tax bill, and time to prepare forms.
   d. The tax preparation firms that charge the least for their service save you the most money.

A **tax table** is a schedule that lists the tax to be paid by a wage earner. This tax is based on total income earned and filing status. Income tax rates are based on the principle of "graduated income tax." This principle means that the amount of tax to be paid increases as income increases.

SCHEDULE I

1998 Tax Table—*Continued*					
If line 6 (Form 1040EZ), line 24 (Form 1040A), or line 39 (Form 1040) is—		**And you are—**			
At least	But less than	Single	Married filing jointly*	Married filing separately	Head of a house-hold
		Your tax is—			
25,000					
25,000	25,050	3,754	3,754	4,254	3,754
25,050	25,100	3,761	3,761	4,268	3,761
25,100	25,150	3,769	3,769	4,282	3,769
25,150	25,200	3,776	3,776	4,296	3,776
25,200	25,250	3,784	3,784	4,310	3,784
25,250	25,300	3,791	3,791	4,324	3,791
25,300	25,350	3,799	3,799	4,338	3,799
25,350	25,400	3,810	3,806	4,352	3,806
25,400	25,450	3,824	3,814	4,366	3,814
25,450	25,500	3,838	3,821	4,380	3,821
25,500	25,550	3,852	3,829	4,394	3,829
25,550	25,600	3,866	3,836	4,408	3,836
25,600	25,650	3,880	3,844	4,422	3,844
25,650	25,700	3,894	3,851	4,436	3,851
25,700	25,750	3,908	3,859	4,450	3,859
25,750	25,800	3,922	3,866	4,464	3,866
25,800	25,850	3,936	3,874	4,478	3,874
25,850	25,900	3,950	3,881	4,492	3,881
25,900	25,950	3,964	3,889	4,506	3,889
25,950	26,000	3,978	3,896	4,520	3,896
* This column must also be used by a qualifying widow(er)					

Answer each question below.

1. The portion of the tax table shown applies to an income equal to or at least $_____ but less than $_____.

2. The income tax paid is based on four types of marital status:

    _____

    _____

    _____

    _____

3. The 1998 Tax Table shown can be used for persons who are single, married, or the head of a household.   True   False

4. The tax paid by a single woman on an income of $25,425 per year is $3,810.   True   False

5. A man who earns $25,950 a year and pays more than $4,000 in taxes falls into the category of "Married, filing separately."   True   False

6. A single person earning $25,000 a year pays _____ less taxes than a single person earning $25,955.

    a. $224   **c.** $220
    b. $194   **d.** $180

7. With an income level of $25,558, a single person pays _____ than a married person filing jointly with a spouse.

    a. $30 more   **c.** $50 less
    b. $60 more   **d.** $60 less

8. Members of the filing category who pay the most tax at every income level shown on Schedule I are

    a. single   **c.** married filing separately
    b. married filing jointly   **d.** head of household

9. All of the following statements can be concluded from the Tax Table, Schedule I, EXCEPT:

    a. As income increases, total taxes paid in every filing category increase.
    b. Taxes paid each year are in part dependent on marital status.
    c. Married individuals filing jointly pay the same amount of taxes as heads of household.
    d. As income decreases, taxes also decrease.
    e. Reducing the tax rate would benefit heads of household more than any other group.

Schedule and Chart Review

Do all the following problems. Work accurately, but do not use outside help. After completing the review, check your answers with the key at the back of the book.

CHART A

UNITED STATES MILEAGE CHART											
For Selected Cities	Atlanta, GA	Chicago, IL	Dallas, TX	Denver, CO	Detroit, MI	Kansas City, MO	Los Angeles, CA	Louisville, KY	Memphis, TN	Milwaukee, WI	Minneapolis, MN
Atlanta, GA		674	795	1398	699	798	2182	382	371	761	1068
Chicago, IL	674		917	996	266	499	2054	292	530	87	405
Dallas, TX	795	917		781	1143	489	1387	819	452	991	936
Denver, CO	1398	996	781		1253	600	1059	1120	1040	1029	841
Detroit, MI	699	266	1143	1253		743	2311	360	713	353	671
Kansas City, MO	798	499	489	600	743		1589	520	451	537	447
Los Angeles, CA	2182	2054	1387	1059	2311	1589		2108	1817	2087	1889
Louisville, KY	382	292	819	1120	360	520	2108		367	379	697
Memphis, TN	371	530	452	1040	713	451	1817	367		612	826
Milwaukee, WI	761	87	991	1029	353	537	2087	379	612		332
Minneapolis, MN	1068	405	936	841	671	447	1889	697	825	332	

Answer each question by filling in the blank, answering true or false, or choosing the best multiple-choice response.

1. The numbers on the chart represent _____.

2. Chart A is used for determining _____ between _____.

3. The mileage between Atlanta and Louisville is 382 miles.   True   False

4. There are 1,059 miles between Los Angeles and Dallas.   True   False

5. The mileage from Chicago to Denver is approximately _____ more than the mileage from Chicago to Memphis.

    a. 500 miles
    b. 200 miles
    c. 400 miles
    d. 700 miles
    e. 600 miles

6. What are the total miles driven on a trip from Dallas to Denver and then from Denver to Kansas City?

    a. 1,237 miles
    b. 1,266 miles
    c. 1,381 miles
    d. 532 miles
    e. 650 miles

7. If a trip from Los Angeles to Chicago took 5 days of driving time, approximately what is the average number of miles driven each day?

    a. 2,000 miles
    b. 300 miles
    c. 100 miles
    d. 500 miles
    e. 400 miles

8. Averaging 400 miles for each tank of gas, how many times would you need to fill the tank as you drove from Kansas City to Atlanta?

    a. 4 times
    b. 2 times
    c. 8 times
    d. 3 times
    e. 6 times

SCHEDULE A

the leisure bus ... **GO TRAVEL–WAYS**								
Scheduled Departures from Bustown, USA								
NORTHBOUND TO					**SOUTHBOUND TO**			
Portland		Seattle			San Francisco		Los Angeles	
A.M.	P.M.	A.M.	P.M.		A.M.	P.M.	A.M.	P.M.
2:05	2:20	1:10	2:15		1:40	5:30	2:40	4:25
5:25	4:50	5:30	4:50		5:20	10:15	8:20	9:30
6:05	5:40	7:25	9:00		11:45		10:50	
7:25	7:10	9:10	11:45					
9:10	7:50	11:35						
11:35	9:00							
	10:00							
	11:45							

Answer each question by filling in the blank, answering true or false, or choosing the best multiple-choice response.

9. Schedule A tells only the times when buses _____ the Bustown, USA, bus depot.

10. The bus departure schedule tells passengers the times when buses leave northbound to the cities of _____ and _____ and southbound to the cities of _____ and _____.

11. Schedule A tells the time when buses arrive at their destinations.   True   False

12. The next bus going to Seattle after 11:45 P.M. leaves at 2:15 P.M.   True   False

13. A passenger who misses the 9:30 P.M. bus to Los Angeles must wait about _____ hours for the next one.

    a. 3
    b. 4
    c. 5
    d. 12
    e. 24

14. The earliest daily bus to Los Angeles leaves the station at

    a. 1:20 A.M.
    b. 6:30 A.M.
    c. 7:10 A.M.
    d. 2:40 A.M.
    e. 3:30 P.M.

15. Which graph most accurately represents Schedule A?

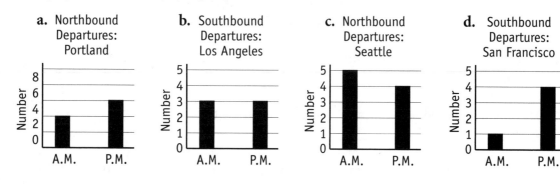

a. Northbound Departures: Portland   **b.** Southbound Departures: Los Angeles   **c.** Northbound Departures: Seattle   **d.** Southbound Departures: San Francisco

16. Seattle is located north of Portland. Many of the departure times listed for Seattle are also listed for Portland. You can infer the following from Schedule A:

    a. One bus goes directly to Portland, while another goes directly to Seattle.
    b. The buses that go northbound to Seattle are the same buses that are going to Portland.
    c. Due to demand, several buses leave the station at the same time.
    d. There are more passengers going to Seattle than to Portland.
    e. There are more passengers going to Portland than to Seattle.

SCHEDULE B

Yearly Interest Schedule Rate: 7%	MONTHLY PAYMENT Amount of Car Loan				
No. of Months	$10,000	$11,000	$12,000	$13,000	$14,000
12	$865.27	$951.80	$1038.33	$1124.85	$1211.38
18	$586.85	$645.54	$704.22	$762.91	$821.59
24	$447.73	$492.50	$537.28	$582.05	$626.82
30	$364.32	$400.76	$437.19	$473.62	$510.05
36	$308.78	$339.65	$370.53	$401.41	$432.28
48	$239.47	$263.41	$287.36	$311.31	$335.25

Answer each question by filling in the blank, answering true or false, or choosing the best multiple-choice response.

17. Schedule B is a payment schedule for a _____ loan borrowed at _____ percent yearly interest rate.

18. The payment schedule applies to loans in the amount of $_____ to $_____.

19. A car loan of $12,000 borrowed for 12 months requires a monthly payment of $1,038.33.   True   False

20. To keep monthly payments below $250 on a $10,000 loan means borrowing the money for 48 months.   True   False

21. For a two-year loan, what is the difference between the monthly payments on a $12,000 loan and a $13,000 loan?

    a. $44.23
    b. $44.77
    c. $36.43
    d. $58.69
    e. $23.95

22. For a $10,000 loan, monthly payments made over a year and a half are _____ higher than monthly payments made over three years.

    a. $55.54
    b. $178.06
    c. $278.07
    d. $261.05
    e. none of the above

23. Which graph most accurately represents Schedule B?

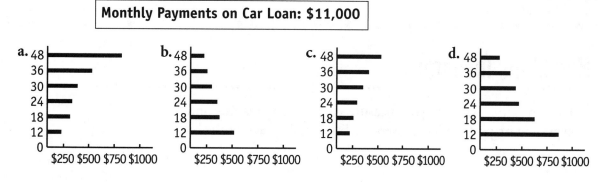

Monthly Payments on Car Loan: $11,000

24. Which statement best describes Schedule B?

a. More loans are made for 48 months than for any other time period.

b. Monthly payments on a car are determined by the amount of the down payment and by credit status.

c. Monthly payments on a car loan are determined by interest rate, amount borrowed, and the amount of time for repayment.

d. High interest rates for car loans are due to inflation.

e. A 48-month loan costs less than a 12-month loan because the monthly payments are less.

Schedule and Chart Review Chart

Circle the number of any problem that you missed and review the appropriate pages. A passing score is 20 correct answers. If you miss more than 4 questions, you should review this chapter.

PROBLEM NUMBERS	SKILL AREA	PRACTICE PAGES
1, 2, 3, 4, 5, 6, 7, 8	chart	80–87, 92–95
9, 10, 11, 12, 13, 14, 15, 16, 17, 18, 19, 20, 21, 22, 23, 24	schedule	80–85, 88–95

MAPS

Map Skills Inventory

This inventory will help you measure your skills in reading and interpreting maps. Correct answers are listed at the back of the book.

MAP A

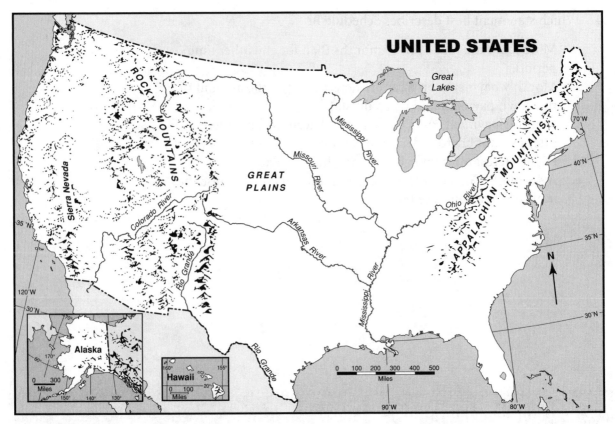

Answer each question below.

1. On Map A, the symbol ![symbol] represents _____.

2. The lines that run across the United States from left to right (labeled 30°, 35°, 40°) are called _____.

3. The Missouri River is in the northeastern United States.   True   False

4. The Rocky Mountains is a major mountain range in the western United States.   True   False

5. How many miles does the Appalachian Mountains stretch?

    a. 1,800   **b.** 2,400   **c.** 1,300   **d.** 300   **e.** 900

MAP B   **Downtown Washington, D.C.**

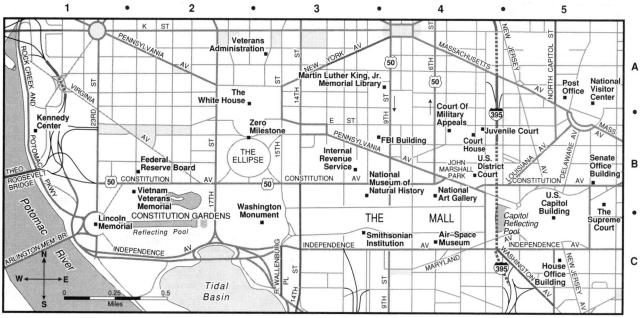

Answer each question by filling in the blank, answering true or false, or choosing the best multiple-choice response. A map-section code is given in parentheses for each place mentioned in the questions.

6. The above map shows the _____ area of Washington, D.C.

7. The Supreme Court (B-5) is located about 0.25 miles _____ of the U.S. Capitol Building (C-5).
   (direction)

8. The Lincoln Memorial is located in map section C-2.   True   False

9. 17th Street crosses Highway 50 less than $\frac{1}{2}$ mile from the   True   False
   Washington Monument (C-2).

10. About how far is the distance between the U.S. Capitol Building
    (C-5) and the White House (A-2)?

    a. $\frac{1}{2}$ mile   **b.** 1 mile   **c.** $1\frac{1}{2}$ miles   **d.** 2 miles   **e.** $2\frac{1}{2}$ miles

11. Starting at the FBI Building (B-3), if you walk southeast on
    Pennsylvania Avenue for 0.5 mile, you end up near what building?

    a. Air-Space Museum   **d.** U.S. District Court
    b. Lincoln Memorial   **e.** U.S. Capitol Building
    c. Internal Revenue Service

MAP C

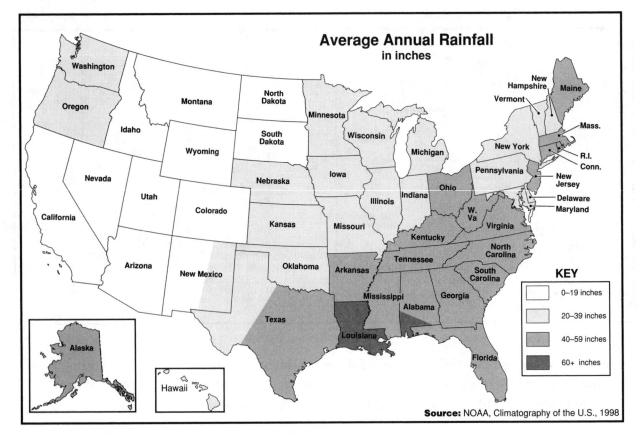

Average Annual Rainfall
in inches

KEY

0–19 inches

20–39 inches

40–59 inches

60+ inches

Source: NOAA, Climatography of the U.S., 1998

Answer each question by filling in the blank, answering true or false, or choosing the best multiple-choice response.

12. The scale of miles measures a total of _____ miles.

13. Map C shows the normal average _____ in _____ for the United States.

14. The symbol [ ] means that the average rainfall for a particular state is from _____ to _____ inches per year.

15. Ohio receives 40–59 inches of rain per year.   True   False

16. Only one state receives two measurably different amounts of rainfall yearly.   True   False

17. Of the following states, which state has the least average rainfall per year?

    a. Hawaii
    b. Texas
    c. Illinois
    d. Wyoming
    e. Alabama

18. Of the following states, which state shown by the map has the driest climate?

    a. Maine
    b. New York
    c. Illinois
    d. California
    e. Nebraska

Map Skills Inventory Chart

Use this inventory to see what you already know about maps and what you need to work on. A passing score is 15 correct answers. Even if you have a passing score, circle the number of any problems that you miss and turn to the practice pages indicated for further instruction.

PROBLEM NUMBERS	SKILL AREA	PRACTICE PAGES
1, 2, 3, 4, 5	geographical map	106–123
6, 7, 8, 9, 10, 11	directional map	106–117, 124–129
12, 13, 14, 15, 16, 17, 18	informational map	106–117, 130–135

What Are Maps?

A **map** is a visual display that represents the whole earth or a particular region of it. Some maps show natural features such as land masses, mountains, rivers, and oceans. Other maps show man-made features such as boundary lines between countries or states and the location of cities and highways. Maps are also used to show special information such as weather conditions and time zones.

Types of Maps

Maps are widely used in education, business, and recreation. Because of their wide variety of uses, it is convenient to group maps into three categories for study in this book: **geographical, directional,** and **informational.**

GEOGRAPHICAL MAPS

A **geographical map** shows the natural features of a region of the earth. **Natural features** are the lands, rivers, lakes, oceans, and other features that are not man-made.

The illustration below is an example of a geographical map. This map of North America shows height (elevation) by darkening or shading mountainous or other elevated land areas.

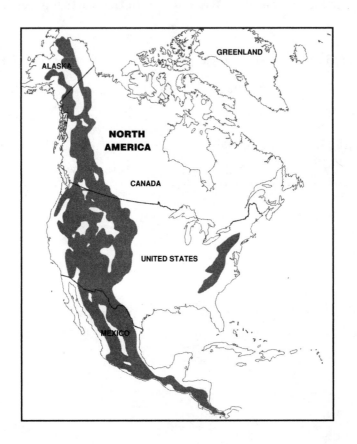

DIRECTIONAL MAPS

A **directional map** is used to show the location of cities, highways, and points of interest. Directional maps called "road maps" are commonly used by travelers to find their way across the country, across a state, or within a city. The use of these maps allows a traveler to plan a route in advance and to estimate time of travel.

The map at the right shows the main highways within the state of Texas.

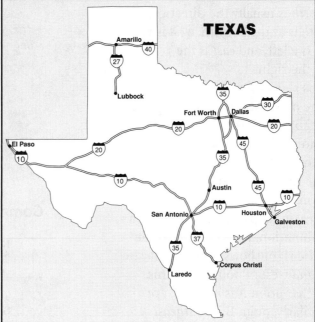

INFORMATIONAL MAPS

An **informational map** gives specific information about a particular area, or it gives information comparing different areas.

This informational map compares the birth rates in the Canadian provinces.

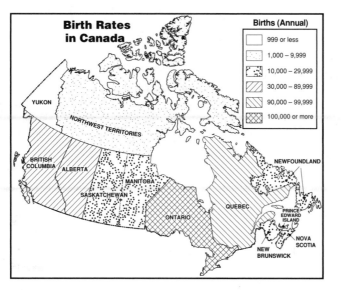

Knowing Directions on a Map

The natural shape of the earth is a sphere (ball-like shape) as shown at the right. A map in the shape of a sphere is called a **globe.** The **North Pole** is at the top of the globe, and the **South Pole** is at the bottom.

When a flat map is drawn of a region of the earth, *north* is usually the direction at the top. *South* is at the bottom, *west* is the direction to the left, and *east* is the direction to the right.

Most maps show direction by a symbol placed on the map itself. Occasionally, the top of the map may be some direction other than north. In this case, the direction symbol will indicate proper directions. Shown at the right are several common map direction symbols.

The four main map directions are often used in combination. For example, as the drawing at the right shows, point A may be both north *and* west of point O. In this case, we say that point A is *northwest* of point O. Similarly, point B is *southeast* of point O.

The example below shows several correct uses of map directions.

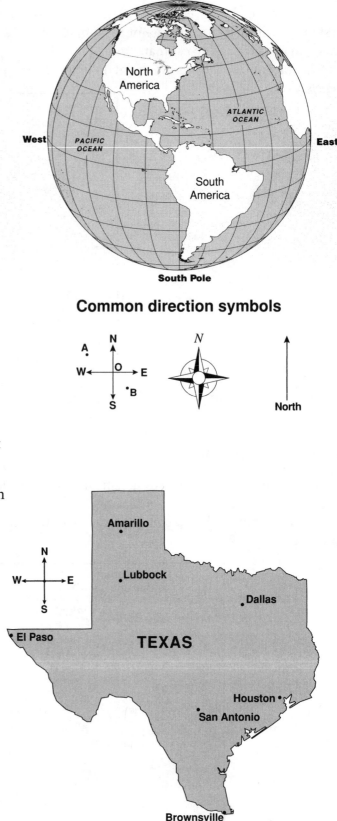

Common direction symbols

EXAMPLE   At the right is a map of Texas with locations of several cities.
- Amarillo is *northwest* of Dallas.
- Brownsville is *southwest* of Houston.
- El Paso is in *western* Texas and is *southwest* of Dallas.
- Houston is in *eastern* Texas.
- San Antonio is *southeast* of Lubbock.

As a quick way for identifying parts of the country, the United States is often divided into the following six regions: Northwest, North Central, Northeast, Southwest, South Central, and Southeast.

Become familiar with these regions before answering the questions below.

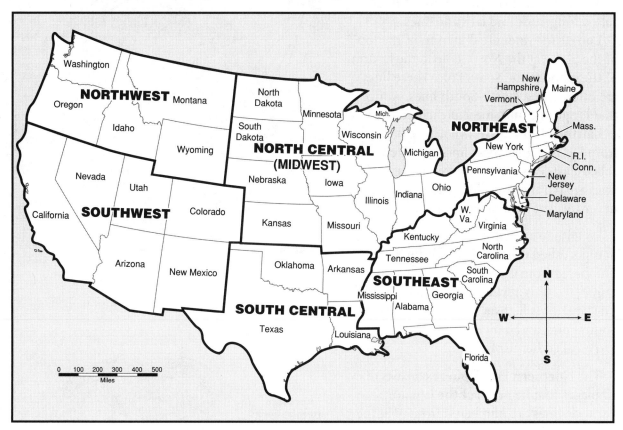

To check your understanding of map directions, fill in each blank below with one of the following directions: north, south, east, west, northeast, northwest, southeast, or southwest.

1. Nevada is _____ of New Mexico.

2. Illinois is _____ of Minnesota.

3. To find Vermont, look _____ of Indiana.

Answer the following questions about the six regions of the United States.

4. The largest region on the West Coast is the _____.

5. The region that contains the least number of states is the _____.

6. Florida is in the region known as the _____.

Understanding Longitude and Latitude Lines

To identify the location of a point on the globe, mapmakers draw two sets of lines called **longitude** and **latitude lines.**

Longitude lines are drawn from the top of the globe at the North Pole to the bottom of the globe at the South Pole. These lines are often called North-South lines. Notice that longitude lines meet at each pole.

Latitude lines are drawn across the globe from left to right. These lines are often called East-West lines. Notice that latitude lines are parallel and circle the earth; they never meet at a point and they never cross.

Longitude and latitude lines are numbered in units called **degrees.** Latitude lines are numbered from 0° to 90°. The latitude line numbered 0° is a special line called the **equator.** The equator is halfway between the North and South Poles, and it divides the globe into two parts called **hemispheres.**

The **Northern Hemisphere** consists of all places that lie north of the equator. Latitude lines are numbered from 0° at the equator to 90° at the North Pole.

The **Southern Hemisphere** consists of all places that lie south of the equator. Latitude lines are numbered from 0° at the equator to 90° at the South Pole.

Longitude lines are numbered from 0° to 180°. The longitude line numbered 0° passes through Greenwich, England. From Greenwich, the longitude lines are numbered from 0° to 180° as you move toward the west, and from 0° to 180° as you move toward the east. The longitude line numbered 180° is exactly halfway around the earth from Greenwich.

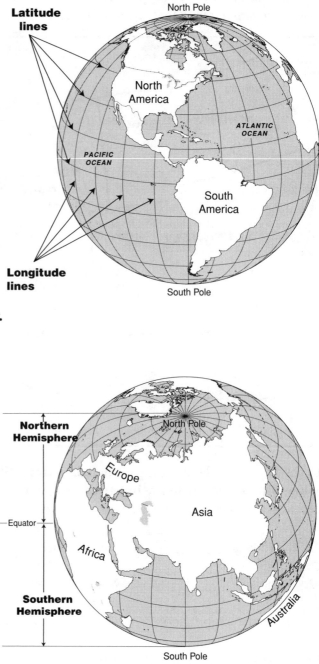

The continents are also often said to be in the **Western Hemisphere** or in the **Eastern Hemisphere.** The globe at the top of the page shows the continents in the Western Hemisphere: North and South America. The globe in the lower half of the page shows the continents of the Eastern Hemisphere: Europe, Asia, Africa, and Australia. The continent of Antarctica is located in the Southern Hemisphere.

All the continents of the earth are shown here on a single flat map. This map is called a "Mercator Projection." Notice that the longitude lines are drawn to be parallel to each other. Although this map shows the exact locations of land and water areas of the earth, it does distort their actual sizes as you look to the far north and far south. In particular, Greenland appears to be much larger on a Mercator Projection than it is on a globe.

Become familiar with this map of the earth and then answer the questions below.

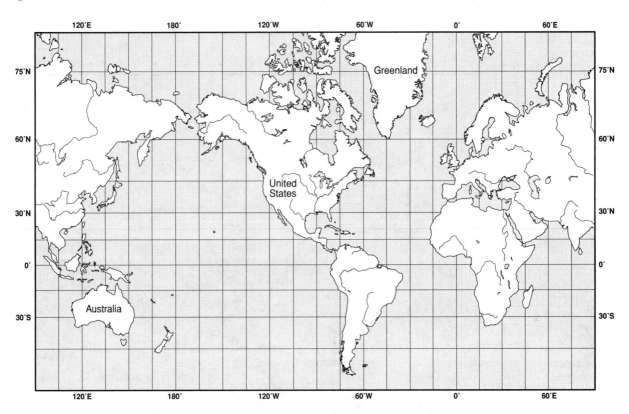

To check your understanding of longitude and latitude lines, fill in each blank below.

1. The lines that run north and south are called _____ lines.

2. The lines that run east and west are called _____ lines.

3. The latitude line that is numbered 0° is called the _____.

4. The latitude line that runs close to the southern border of the United States is the _____ degree line.

5. Australia lies completely _____ of the equator.
   (direction)

Using an Index Guide

On most maps, an **index guide** is used to locate countries, cities, and other points of interest. While longitude and latitude lines identify a point's location on any globe or map, an index guide is a listing that identifies a small area on a map by which a specific place is located. An index guide consists of

1. a row of numbers across the top (or bottom) of the map,
2. a column of letters along the left (or right) side of the map, and
3. a listing of places with a letter-number location code.

COUNTRIES	
Albania	G-6
Austria	F-5
Belarus	D-7
Belgium	E-4
Bosnia	F-6
Bulgaria	G-7
Croatia	F-6
Czech Republic	E-6
Denmark	D-5
Estonia	C-7
Finland	B-7
France	F-3
Germany	E-5
Greece	G-7
Hungary	F-6
Ireland	D-2
Italy	G-5
Latvia	C-7
Lithuania	D-7
Macedonia	G-7
Netherlands	D-4
Norway	B-5
Poland	D-6
Portugal	G-1
Romania	F-7
Slovenia	F-5
Spain	G-2
Sweden	B-6
Switzerland	F-4
United Kingdom	D-3
Yugoslavia	G-6

MAJOR CITIES	
Amsterdam (Neth.)	D-4
Athens (Gr.)	H-7
Barcelona (Sp.)	G-3
Berlin (Ger.)	D-5
Bonn (Ger.)	E-4
Brussels (Bel.)	E-4
Copenhagen (Den.)	D-5
Dublin (Ire.)	D-2
Hamburg (Ger.)	D-5
Helsinki (Fin.)	C-7
Lisbon (Por.)	G-1
Liverpool (U.K.)	D-3
London (U.K.)	D-3
Lyon (Fr.)	F-4
Madrid (Sp.)	G-2
Marseille (Fr.)	F-4
Milan (It.)	F-4
Munich (Ger.)	E-5
Oslo (Nor.)	C-5
Paris (Fr.)	E-3
Rome (It.)	G-5
Stockholm (Swe.)	C-6
Vienna (Aus.)	E-6
Zurich (Swi.)	F-4

MAJOR RIVERS	
Danube	E-5, F-6
Ebro	F-2
Elbe	D-5
Loire	E-3
Po	F-4
Rhine	E-4
Rhone	F-4
Seine	E-3

The example below shows the correct use of an index guide.

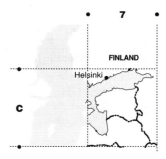

EXAMPLE Locate the city of Helsinki.

> **STEP 1** Find the name of the city Helsinki on the index guide.
> Read the letter-number code: C-7

> **STEP 2** Locate C on the left side of the map. Scan directly to the right, and stop below the number 7. The point of intersection of the two lines of sight (across from C and below 7) gives the area in which Helsinki is located.

ANSWER: Helsinki is on the southern coast of Finland.

Notice that the letters and numbers are centered in the area they refer to. For instance C refers to the area halfway up to B and halfway down to D. The boundaries are marked with a dot.

••

Check your understanding of the use of an index guide. Use the map on the previous page to answer each question below by filling in the blank or by choosing true or false.

1. Write the letter-number code of each of the following cities:

    a. London _____
    b. Athens _____
    c. Hamberg _____

2. The Rhine River is in the country of _____.

3. The large body of water in map section C-6 is the _____.

4. Near what large city does the Elbe River run into the North Sea?
   _____

5. In what country is Lisbon? _____

6. The largest country that shares a border with Poland is Germany.   True   False

7. Stockholm is north of Oslo.   True   False

8. Marseille is in map section F-2.   True   False

Using a Distance Scale

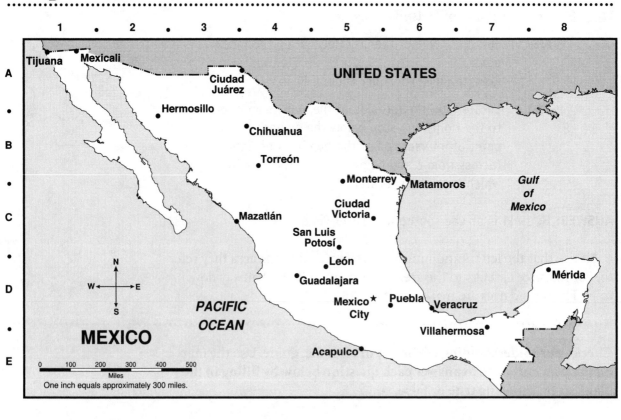

CITIES AND TOWNS

Acapulco	E-5	Hermosillo	B-2	Mexico City	D-5	Tijuana	A-1
Chihuahua	B-4	León	D-5	Monterrey	C-5	Torreón	B-4
Ciudad Juárez	A-4	Matamoros	C-6	Puebla	D-6	Veracruz	D-6
Ciudad Victoria	C-5	Mazatlán	C-3	San Luis Potosí	D-5	Villahermosa	E-7
Guadalajara	D-4	Mérida	D-8				

EXAMPLE   What is the direct distance between Torreón and Matamoros?

STEP 1   Use the index guide to find Torreón and Matamoros.

STEP 2   Use a ruler to measure the approximate distance between them: about $1\frac{1}{2}$ inches.

STEP 3   Multiply $1\frac{1}{2}$ by 300, the number of miles per inch given by the distance formula (found under the map title).

$300 \times 1\frac{1}{2} = 300 \times \frac{3}{2} =$ **450 miles**

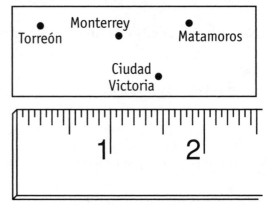

ANSWER: About 450 miles

As the previous example shows, you can use a ruler to measure the **direct distance** between two points. However, the direct distance is not the **road distance.** The actual road distance between two points is always greater than the direct distance measured by a ruler. In the previous example, the actual road distance between Torreón and Matamoros is about 575 miles. Actual road mileage between two cities is often written on the map itself beside the road connecting the cities.

When you don't have a ruler available, you can use a second method to measure direct distance. You can make a **mileage ruler** by drawing a distance scale on a piece of paper. The example below shows this second method.

EXAMPLE   Find the direct distance between the Mexico City and Mérida.

STEP 1   Use the index guide to find the two cities: Mexico City and Mérida.

STEP 2   Use the distance scale to make a mileage ruler. Then measure the direct mileage between the cities. As seen at the right, the total mileage between the cities is about **600 miles.**

ANSWER: About 600 miles.

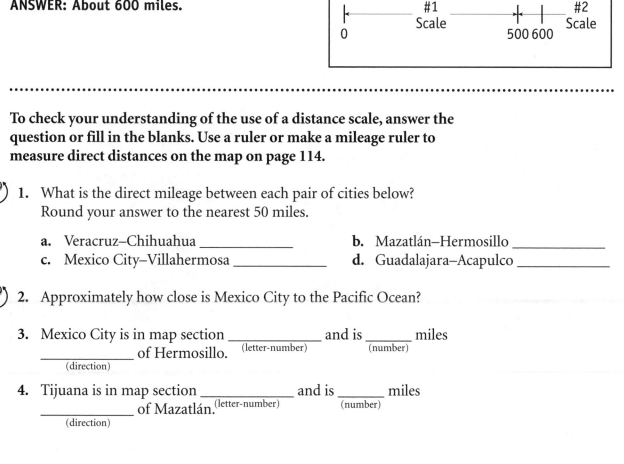

. . . .

To check your understanding of the use of a distance scale, answer the question or fill in the blanks. Use a ruler or make a mileage ruler to measure direct distances on the map on page 114.

1. What is the direct mileage between each pair of cities below? Round your answer to the nearest 50 miles.

    a. Veracruz–Chihuahua _____   **b.** Mazatlán–Hermosillo _____
    c. Mexico City–Villahermosa _____   **d.** Guadalajara–Acapulco _____

2. Approximately how close is Mexico City to the Pacific Ocean?

3. Mexico City is in map section _____ and is _____ miles
    _____ of Hermosillo.   (letter-number)   (number)
    (direction)

4. Tijuana is in map section _____ and is _____ miles
    _____ of Mazatlán. (letter-number)   (number)
    (direction)

Types of Questions

In this next section, you will be working with each of the three types of maps. First, though, look at the three types of questions you'll be asked in your study of maps. These questions will help you find and interpret information on each map.

Scanning the Map Questions

"Scanning the Map" questions require you to be familiar with general features concerning correct use of the map. Answer these questions by filling in words to complete a sentence. To scan a map:

- Pay attention to the title.
- Look for any map symbols, distance scales, index guides, special keys, and any other information located on or around the outside of the map.
- Be familiar with boundary lines and other special names and symbols appearing on the map itself.

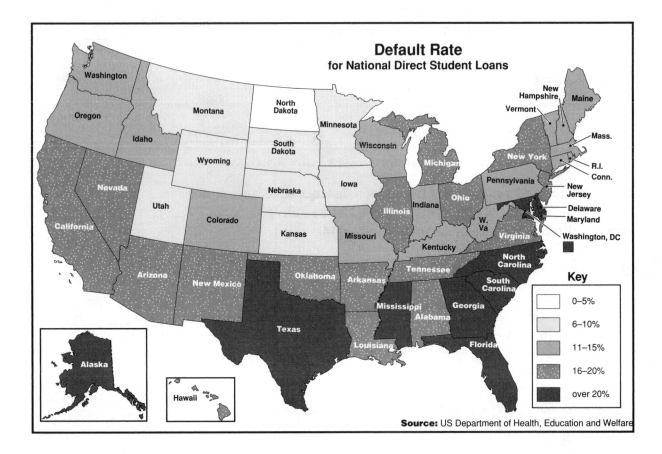

Default Rate
for National Direct Student Loans

Key
- 0–5%
- 6–10%
- 11–15%
- 16–20%
- over 20%

Source: US Department of Health, Education and Welfare

<u>EXAMPLE 1</u>   The above map shows the state default rate for _____.

ANSWER: National Direct Student Loans (found in the title)

EXAMPLE 2 The maximum default rate shown in any state
is _____ .

ANSWER: **over 20%** (found by looking at the listing of symbols in the key
at the right of the map)

Reading the Map Questions

"Reading the Map" questions require you to locate and interpret
information on the map itself. These questions are answered true or false.

EXAMPLE 1 On loans issued to students, Montana has a default rate   True   False
somewhere between 6% and 10%.

ANSWER: **True.** Montana is shaded with light gray. This symbol represents
a rate of 6% to 10%.

EXAMPLE 2 Two states have a default rate in the 0–5% category.   True   False

ANSWER: **False.** The symbol for 0–5% is white (no shading). Only one
state, North Dakota, displays this symbol.

Comprehension Questions

Comprehension questions require you to compare values, make
inferences, and draw conclusions. Answer each question by choosing
the best answer.

EXAMPLE 1 Oregon has a default rate that is _____ Utah's.

    a. the same as   **d.** four times
    b. less than   **e.** one-half
    c. greater than

ANSWER: **c. greater than.** Oregon has a default rate of 11–15%, which is
greater than Utah's rate of 6–10%, but not four times greater
(choice d).

EXAMPLE 2 The number of states that have a default of over 20% is

    a. 5   **d.** 10
    b. 7   **e.** 16
    c. 9

ANSWER: **c. 9.** Count only the states marked with the shading ▇ .
These states are Alaska, Texas, Mississippi, Florida, Georgia,
South Carolina, North Carolina, Maryland, and Delaware. Be
careful not to count Washington, D.C., the nation's capital. The
question asks only for states.

Geographical Maps

A geographical map shows land and water formations. Often, but not always, boundary lines (borders) of countries or states and the locations of cities are also shown. Below are examples of the most commonly used geographical map symbols.

Boundary Lines show the borders or limits of an area.

Rivers show large natural streams of water.

Cities • show population centers.

Mountains show mountain ranges.

Lakes show bodies of water surrounded by land.

State Capitals ★ show the capital of each state.

On many geographical maps, color is used to distinguish areas of different height. Land height (elevation) means "distance above sea level." **Sea level** is the level of the ocean. Thus, land that is the same level as the ocean is said to have 0 elevation, while land that is 500 feet higher than the ocean is said to have an elevation of 500 feet. The map below is an example of a geographical map.

To answer questions about geographical maps, follow the sequence below.

Scanning the Map

To scan a geographical map, identify the title, symbols, elevation key, distance scale, and longitude and latitude lines. Not all maps will have all of these items, but you should be aware of what you are working with.

EXAMPLE   The map on the previous page includes the region called the _____.

ANSWER: Southeastern United States (found in the title)

Reading the Map

To read a geographical map, find the place or area asked about. Interpret symbols and shading, and identify longitude and latitude as needed.

EXAMPLE   Montgomery is the capital of Mississippi.   True   False

STEP 1   Look at the symbols for cities and decide which symbol stands for a state capital. Remember, there is only one state capital in a state.

STEP 2   Find the capital of Mississippi.

ANSWER: False. The capital of Mississippi is Jackson. Montgomery is the capital of Alabama.

Comprehension Questions

Comprehension questions require you to compare values, make inferences, or draw conclusions.

EXAMPLE   What state would be the most likely choice for a mountain vacation?

a. Mississippi
b. Alabama
c. North Carolina
d. South Carolina
e. Florida

STEP 1   Look at each of the states listed above.

STEP 2   Use the map symbol for mountainous areas to decide which state contains the largest amount of mountainous terrain.

ANSWER: c. North Carolina

Practice Geographical Map

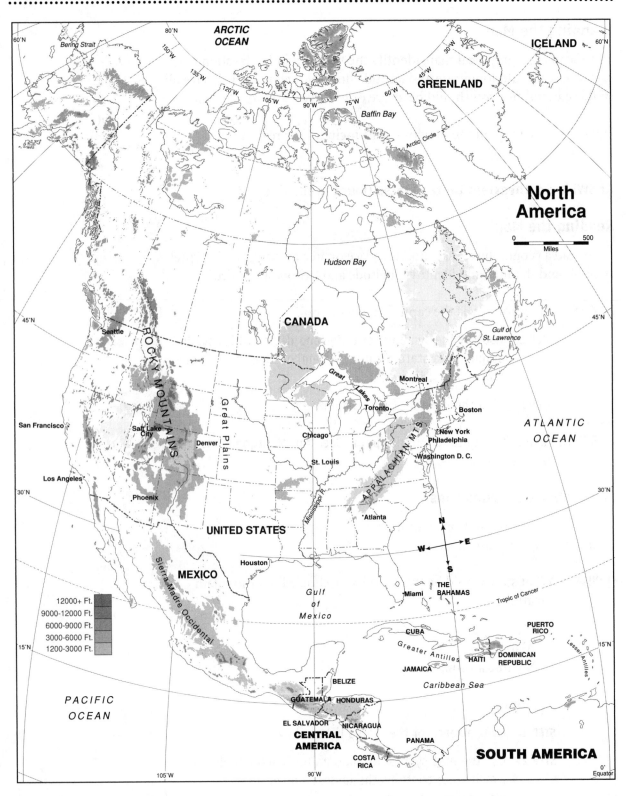

Scanning the Map

1. The dashed line (----) between 60° and 80° latitude stands for a special latitude line called the _____.

2. The map distance scale represents a total distance of _____ miles.

3. The two largest countries on the continent of North America are
   _____ and _____.

4. The _____ longitude line passes through the eastern edge
   of Cuba.

Reading the Map

Decide whether each sentence is true or false and circle your answer.

5. The United States is north of the 20° latitude line.   True   False

6. Hudson Bay is located in Canada.   True   False

7. The equator passes through the continent of North America.   True   False

Comprehension Questions

Answer each question by choosing the best multiple-choice response.

8. You can conclude from the map that the United States has _____
   land area than Mexico.

   a. less   **c.** more
   b. less populated   **d.** more populated

9. Approximately how many miles is it from Los Angeles, California,
   to New York City?

   a. 1,500   **d.** 3,200
   b. 2,000   **e.** 5,000
   c. 2,400

10. From the map of North America, what can you conclude?

   a. North America is characterized by a uniform distribution of
          mountainous areas.
   b. All of the major rivers of North America eventually drain into
          the Atlantic Ocean.
   c. North America contains more natural resources than the
          continent of South America.
   d. North America is the largest of all of the continents north of
          the equator.
   e. North America is characterized primarily by many high
          mountainous areas in the West and by a few low mountain
          ranges in the East.

Geographical Maps: Applying Your Skills

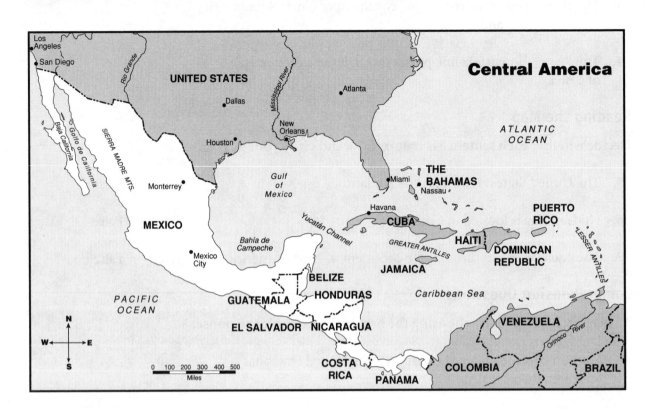

Answer each question by filling in the blank, answering true or false, or choosing the correct multiple-choice response.

1. El Salvador shares borders with the two countries: _____ and _____.

2. Mexico shares borders with two countries to the south: _____ and _____.

3. The largest island off the southern coast of Florida is _____.

4. Puerto Rico is located _____ of the Dominican Republic.
   (direction)

5. The countries of Haiti and the Dominican Republic are part   True   False
   of the same island.

6. Panama shares borders with Costa Rica and Colombia.   True   False

7. Havana, Cuba, is about how many miles from Miami, Florida?

    a. 50
    b. 250
    c. 450
    d. 650
    e. 850

8. Cuba is approximately how many miles from Nicaragua?

    a. 300
    b. 400
    c. 900
    d. 1,200
    e. 1,500

9. Honduras shares borders with the countries of Guatemala, _____ and _____.

    a. Belize, El Salvador
    b. Belize, Nicaragua
    c. El Salvador, Nicaragua
    d. El Salvador, Panama
    e. Panama, Nicaragua

10. To make an overland trip from Nicaragua to El Salvador requires going through what country?

    a. Belize
    b. Costa Rica
    c. Guatemala
    d. Panama
    e. Honduras

11. Looking carefully at the map, what can you conclude?

    a. Central American countries are characterized by a uniform distribution of mountainous areas.
    b. Central American countries are characterized by mountainous interior regions and narrow lowland coastal regions.
    c. Central American countries are surrounded by three large bodies of water.
    d. All of the capitals are located along the Pacific coast.
    e. The fastest travel route is from north to south.

Directional Maps

You are probably very familiar with directional maps. City maps are used to locate streets, buildings of interest, and points of interest such as parks. State and country maps are used to show the location of cities and of the highway systems that connect them. Before you work with directional maps, look at some commonly used symbols that appear on them.

Roads and highways are shown by thick or thin lines.

——— Two-lane roads

——— Divided highway, four lanes or more

═══ Usually an interstate or U.S. highway

Cities are shown by dots and circles.

● Cities

★ State capitals

Highway route numbers are shown as numbers inside special symbols.

⑦⑤ Interstate highway 75

⑤⑨ U.S. highway 59

⑫ State highway 12

Other symbols

✈ Airport

═╪═ Intersection of roads and major highways

The map below is a section of a directional map. Refer to this map as you read the example questions on the next page.

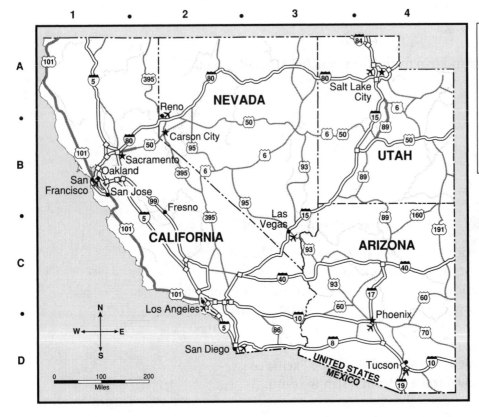

Cities	
Las Vegas	C-3
Oakland	B-1
Phoenix	D-4
Reno	B-2
Sacramento	B-1
Salt Lake City	A-4
San Diego	D-2
San Francisco	B-1
San Jose	B-1
Tucson	D-4

To answer questions about directional maps, follow the sequence below.

Scanning the Map

To scan a directional map, notice the map title, index guide, direction key, and distance scale.

EXAMPLE   The map distance scale represents a total distance of
_____ miles.

ANSWER: 200 miles

Reading the Map

To read a directional map, use the index guide to locate places, and use the direction key and distance scale to identify directions and compute distances.

EXAMPLE   The major interstate highway running between Phoenix and   True   False
Los Angeles is (17) .

STEP 1   Use the index guide to find both cities.

STEP 2   Identify the interstate highway running between them.

ANSWER: False. Interstate (10) runs between Phoenix and Los Angeles. Notice that (17) runs north and south out of Phoenix.

Comprehension Questions

To answer comprehension questions about directional maps, compare locations and distances, make inferences, and draw conclusions.

EXAMPLE   The direct distance between Sacramento and Los Angeles is about _____ the direct distance between Sacramento and Reno.

   a. 250 miles less than   **d.** 220 miles more than
   b. 100 miles less than   **e.** 100 miles more than
   c. the same as

STEP 1   Measure the approximate distance between Sacramento and each other city.
Sacramento to Los Angeles is about 350 miles.
Sacramento to Reno is about 130 miles.

STEP 2   Subtract the distances.
350 – 130 = 220

ANSWER: d. 220 miles more than

Practice Directional Map

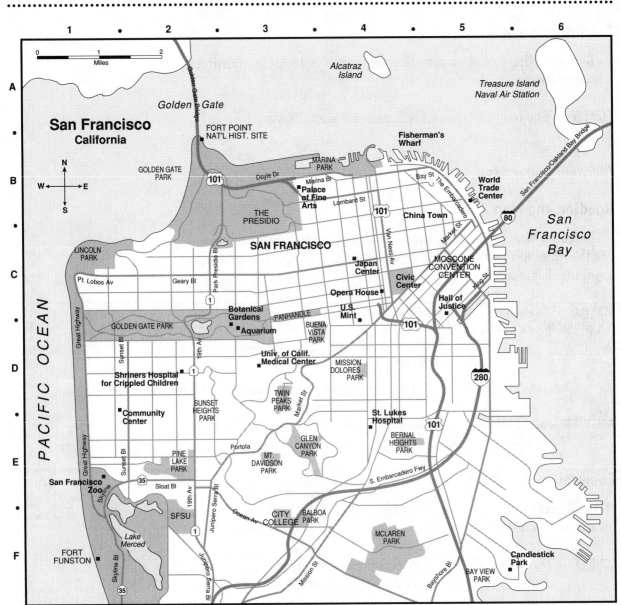

POINTS OF INTEREST

Alcatraz Island	A-4
Buena Vista Park	D-3
Candlestick Park	F-5
China Town	B-4
Civic Center	C-4
Community Center	D-1
Fisherman's Wharf	B-4
Glen Canyon Park	E-3
Golden Gate Bridge	A-2
Golden Gate Park	B-2
Hall of Justice	C-5
Japan Center	C-4
Lake Merced	F-2
Lincoln Park	C-1
Marina Park	B-4
Mt. Davidson Park	E-3
Opera House	C-4
Palace of Fine Arts	B-3
Pine Lake Park	E-2
San Francisco Zoo	E-1
Sunset Heights Park	D-2
U.S. Mint	C-4
World Trade Center	B-5

Scanning the Map

Fill in each blank as indicated.

1. The map shows points of interest and street names in the city of _____.

2. According to the index guide, Alcatraz Island is located in the map section _____.
   (letter-number)

3. San Francisco Bay surrounds San Francisco on both the north and _____ sides.
   (direction)

Reading the Map

Decide whether each sentence is true or false and circle your answer.

4. Interstate 180 is also called the Southern Embarcadero Freeway where it passes through the city of San Francisco. True False

5. Candlestick Park is more than 1 mile from San Francisco Bay. True False

6. The Japan Center is near Van Ness Avenue and Geary Boulevard. True False

7. Another name for 19th Avenue is Park Presido Blvd. True False

Comprehension Questions

Answer each question by choosing the best multiple-choice response.

8. The Botanical Gardens (C-2) are located in

   a. Buena Vista Park **d.** Candlestick Park
   b. Glen Canyon Park **e.** Mt. Davidson Park
   c. Golden Gate Park

9. The walking distance from the Japan Center to Fisherman's Wharf is about

   a. 1 mile **d.** 4 miles
   b. 2 miles **e.** 5 miles
   c. 3 miles

10. The directional map of San Francisco contains the following information:

   a. location and names of all streets
   b. points of interest and major highway numbers
   c. land and water elevation

Directional Maps: Applying Your Skills

CITIES AND TOWNS	
Aurora	B-4
Bloomington	D-3
Carbondale	G-3
Champaign	D-4
Chicago	B-5
Chicago Heights	B-5
Danville	D-5
Decatur	D-4
East St. Louis	F-2
Elgin	A-4
Evanston	A-5
Freeport	A-3
Galesburg	C-2
Highland Park	A-5
Joliet	B-4
Kankakee	C-5
Normal	D-3
Peoria	C-3
Rockford	A-3
Skokie	A-5
Springfield	E-3
Urbana	D-4
Waukegan	A-5

Answer each question by filling in the blank, answering true or false, or choosing the correct multiple-choice response.

1. According to the index guide, the city of Elgin is located in the map section _____.
   _(letter-number)

2. The capital city of Illinois is _____.

3. The large body of water that Chicago is located near is _____.

4. Decatur is about 40 miles west of Springfield.　　　　True　　　False

5. The direct distance by air between Peoria and Chicago is about 125 miles.　　　　True　　　False

6. The major interstate approaching Chicago from the south is 90 .　　　True　　　False

7. The driving distance between East St. Louis and Chicago is just a little less than

   a. 100 miles　　　　d. 400 miles
   b. 200 miles　　　　e. 500 miles
   c. 300 miles

8. If you leave Chicago driving south on 57 , and then turn east on 74 , what city will you come to?

   a. Danville　　　　d. East St. Louis
   b. Normal　　　　e. Rockford
   c. Peoria

9. Urbana is about the same distance from Chicago as it is from the city of

   a. Kankakee　　　　d. Rockford
   b. Decatur　　　　e. Galesburg
   c. East St. Louis

10. At 55 miles per hour, what is the approximate driving time between Urbana and Chicago? (**Remember:** To estimate driving time, divide the distance traveled by the average speed.)

    a. $\frac{1}{2}$ hour　　　　d. $3\frac{1}{2}$ hours
    b. 1 hour　　　　e. 5 hours
    c. 2 hours

Informational Maps

An informational map is the most common type of map that appears in newspapers and magazines. Unlike geographical and directional maps, each informational map shows a special kind of information. Examples of informational maps include weather maps, zip code maps, time zone maps, and maps dealing with such concerns as unemployment. You will be working with several of these examples in this section.

One characteristic of informational maps is that each may have its own set of unique symbols. As an example, look at the weather forecast map below.

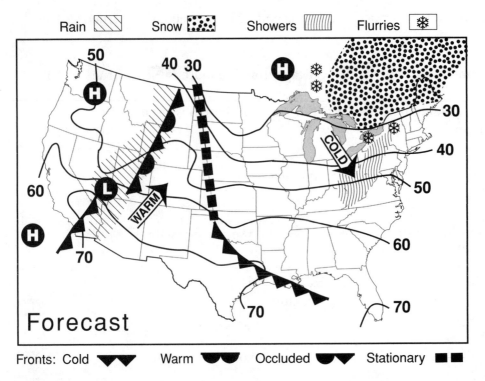

Several symbols are defined on the map, and several are defined below.

Symbols Defined on the Map

Snow

Flurries: light snow and wind

Rain

Showers: brief periods of rain

Definitions of Other Symbols

Low atmospheric pressure. A "low" indicates moist air and cool, wet weather. **L**

High atmospheric pressure. A "high" indicates dry air and warm, dry weather. **H**

(**Note:** Temperatures are indicated by numbers around the border of the U.S. The lines connecting the same temperature readings indicate locations with the same temperature. These lines are called "constant temperature lines" or **isotherms.**)

To answer questions about informational maps, follow the sequence below.

Scanning the Map

To scan an informational map, notice the graph title and any special symbols used on the map.

EXAMPLE   The symbol ❄ means _____.

ANSWER: snow flurries

Reading the Map

To read an informational map, identify map symbols and other data on the map itself. Then locate the specific information needed.

EXAMPLE   The high temperatures shown for most of the southeastern   True   False
United States are in the 40s.

STEP 1   Locate the southeastern states in the lower right-hand corner of the country.

STEP 2   Read the temperatures indicated.

ANSWER: False. The high temperatures in the southeastern states are in the 50s, 60s, and, to a small extent, the 70s.

Comprehension Questions

To answer comprehension questions, compare the use of map symbols and other data on the map.

EXAMPLE   This could be a weather map during the month of

a. January
b. May
c. June
d. July
e. August

STEP 1   Notice that the moisture shown includes snow and flurries.

STEP 2   Combine this with the information that the temperatures range from the 30s to the 70s.

ANSWER: a. January

Practice Informational Map

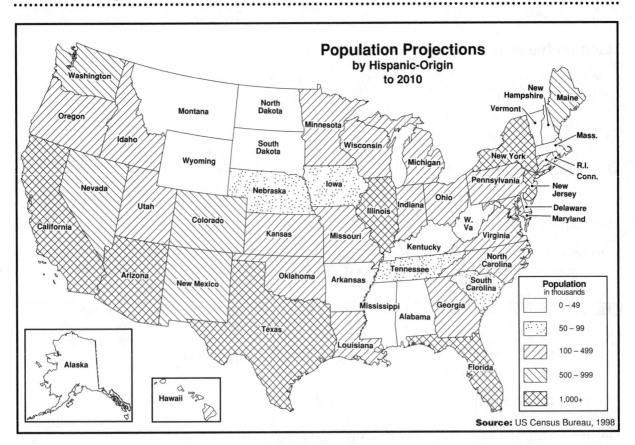

Population Projections
by Hispanic-Origin
to 2010

Population
in thousands

	0 – 49
	50 – 99
	100 – 499
	500 – 999
	1,000+

Source: US Census Bureau, 1998

Scanning the Map

Fill in each blank as indicated.

1. The map above shows the population projected to the year
   _____ for persons of _____ origin.

2. In the western part of the United States, the Hispanic population in
   the states of _____ and _____ are expected to
   reach one million by 2010.

3. The symbol 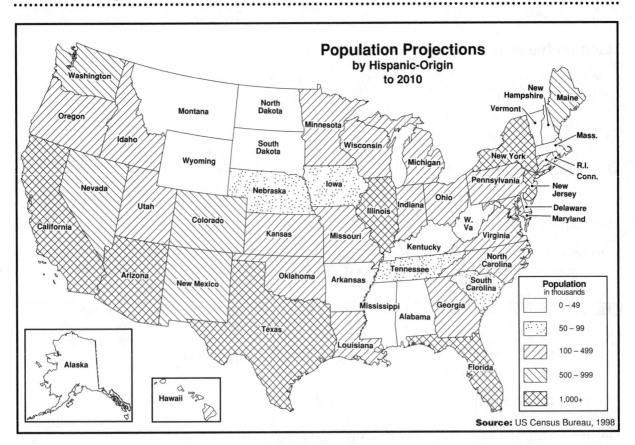 stands for approximately _____ to
   _____ thousand persons on this map.

4. The information collected for this map was obtained from the
   _____.

READING THE MAP

Decide whether each sentence is true or false and circle your answer.

5. The southern state projected to have the largest Hispanic population to 2010 is Georgia.

    True   False

6. The number of Hispanic persons in Colorado is projected to be greater than Wyoming and Montana combined.

    True   False

7. The projections of Hispanic persons in New Mexico will be greater than in the state of New York.

    True   False

Comprehension Questions

Answer each question by choosing the best multiple-choice response.

8. The projection of persons of Hispanic origin in Utah is expected to be more than what state below?

    a. Idaho
    b. Texas
    c. North Carolina
    d. Arkansas
    e. Washington

9. Of the southern states, which state is projected to have the smallest Hispanic population in 2010?

    a. Alabama
    b. Georgia
    c. South Carolina
    d. North Carolina
    e. Florida

10. All of the following statements can be concluded from this map EXCEPT:

    a. Some of the largest Hispanic populations are projected to be in the southwestern portion of the U.S.
    b. The smallest Hispanic populations are projected to be in the north central states.
    c. The two largest states are expected to have the largest projections of Hispanics.
    d. The northeastern states are expected to have the smallest Hispanic populations through 2010.
    e. None of the northwestern states are projected to have more Hispanics than California.

Informational Maps: Applying Your Skills

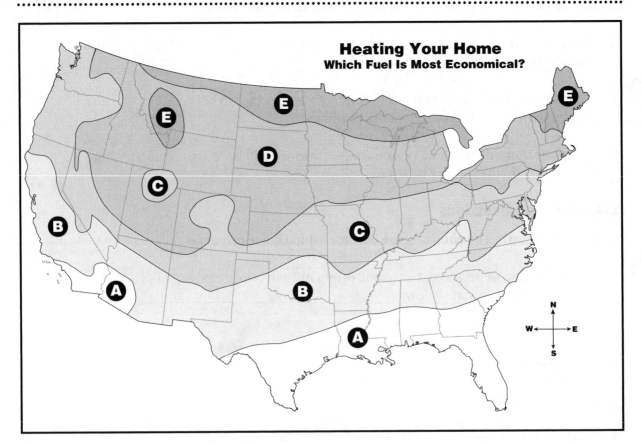

Heating Your Home
Which Fuel Is Most Economical?

Average Annual Home Heating Costs with Alternate Fuels*

*Based on a 1,200 square-foot home with average insulation and weatherization. Local climate and fuel price difference will cause heating costs to vary.

Fuel & Costs	Zone A	B	C	D	E
Oil	275	687	963	1155	1375
Gas	193	484	550	754	1001
Electricity	440	1089	1529	1848	2173
Wood	180	414	624	744	876

Answer each question by filling in the blank, answering true or false, or choosing the correct multiple-choice response.

1. The table and map above compare the average annual
   _____ costs for each of _____ zones or
   (number)
   areas of the country.

2. In every zone, the cost of _____ is higher than any other
   listed fuel.

3. The zone showing the lowest average cost for each of the alternative fuels is _____.

4. Residents of Zone B who heat their home with gas can expect an average cost of $687 per year.　　　True　　　False

5. Residents of Zone B who use oil pay higher heating costs than residents of Zone C who use wood.　　　True　　　False

6. In Zone E it costs twice as much as in Zone _____ to heat with gas.

   a. A　　　　　　　　　c. C
   b. B　　　　　　　　　d. D

7. According to the table and map, states in Zone D pay four times as much for electricity as states in

   a. Zone A　　　　　　　d. Zone D
   b. Zone B　　　　　　　e. Zone E
   c. Zone C

8. The cost per year of heating with gas in Zone B is approximately _____ less than in Zone E.

   a. $300　　　　　　　　d. $200
   b. $500　　　　　　　　e. $100
   c. $700

9. In Zone D, the cost per year of heating with electricity is approximately _____ more than the cost of heating with oil.

   a. $500　　　　　　　　d. $900
   b. $600　　　　　　　　e. $1,000
   c. $800

10. Which statement best summarizes the information on this table and map?

   a. Heating costs are consistently higher in states on the West Coast than in states on the East Coast.
   b. Electricity will continue to be the most expensive source of heating costs in future years.
   c. Oil and gas used as heating fuels are more expensive in southern states than in northern states.
   d. Although costs vary from zone to zone, heating costs are consistently lower in southern states than in northern states.
   e. Many people are changing to electric heat from coal because electricity is an environmentally cleaner source of energy.

Map Review

To find out how well you understand maps, do the following problems. Answer carefully, but do not use outside help. When you've finished, check your answers with the key at the back of the book.

MAP A

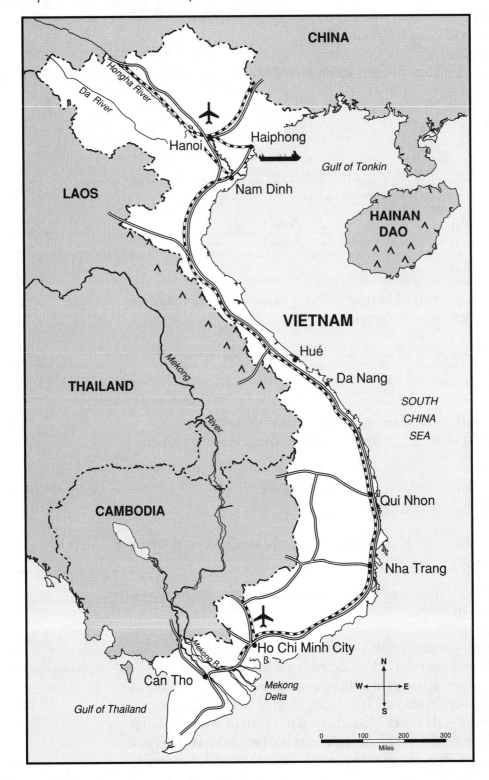

Answer each question by filling in the blank, answering true or false, or choosing the best multiple-choice answer.

1. The distance scale measures _____ miles.

2. The _____ River flows through Cambodia.

3. Ho Chi Minh City is located in southern Laos.   True   False

4. The island of Hainan Dao lies approximately 250 miles   True   False
   off the coast of Vietnam.

5. According to the map, both _____ and _____ have
   mountainous areas.

   a. Northern Laos and Cambodia
   b. Southern Laos and Hainon Dao
   c. Cambodia and Thailand
   d. Laos and Thailand
   e. Southern Laos and Cambodia

6. Laos and Vietnam are separated by

   a. the Gulf of Tonkin
   b. the Hongha River
   c. the Hanoi airport
   d. a mountain range
   e. Thailand

7. Approximately what is the distance from Can Tho to
   Ho Chi Minh City?

   a. 150 miles
   b. 200 miles
   c. 600 miles
   d. 3 miles
   e. 10 miles

8. The road between Haiphong harbor and Nam Dinh

   a. is longer than the Da River
   b. crosses a mountain range
   c. is approximately the same length as the road between Hue and
      Da Nang
   d. crosses the Da River
   e. crosses the Nam Dinh harbor

MAP B

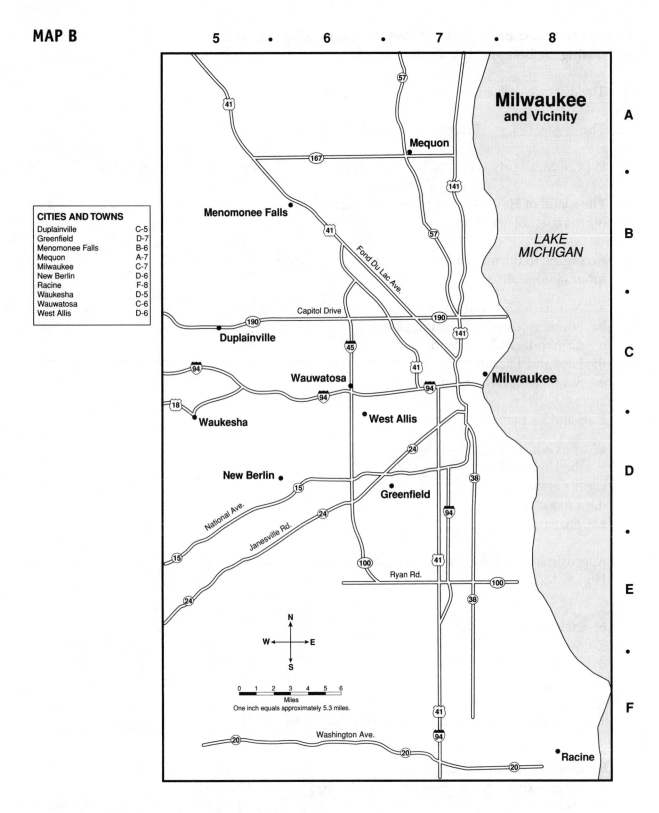

Answer each question by filling in the blank, answering true or false, or choosing the best multiple-choice answer.

9. On the distance scale, 1 inch represents a distance of
approximately _____ miles.
(number)

10. The index guide indicates that Mequon is located in the section

    _____.
    (letter-number)

11. Menomonee Falls is located southwest of Milwaukee.   True   False

12. Driving at 60 miles per hour, you would need to plan for about   True   False
    $1\frac{1}{2}$ hours of driving time for a trip from Milwaukee to Chicago
    (a distance of 90 miles).

13. If you drive west from Milwaukee on 🛡94, turn north on 🛡45, and
    then turn west on ⑲⓪, you will reach the town of _____.

    a. Wauwatosa
    b. Waukesha
    c. New Berlin
    d. Duplainville
    e. Greenfield

14. From this map, you can tell that Milwaukee is located near a

    a. state capital
    b. lake
    c. ocean
    d. mountain range
    e. plateau

15. Highway Route ⑲⓪ is also named

    a. Fond Du Lac Avenue
    b. Ryan Road
    c. Capitol Drive
    d. National Avenue
    e. Washington Avenue

16. Traveling north on Route ㊶, the distance between the Route ⑳
    turnoff and the Route ⑩⓪ turnoff is about how many miles?

    a. 5
    b. 10
    c. 50
    d. 80
    e. 100

MAP C

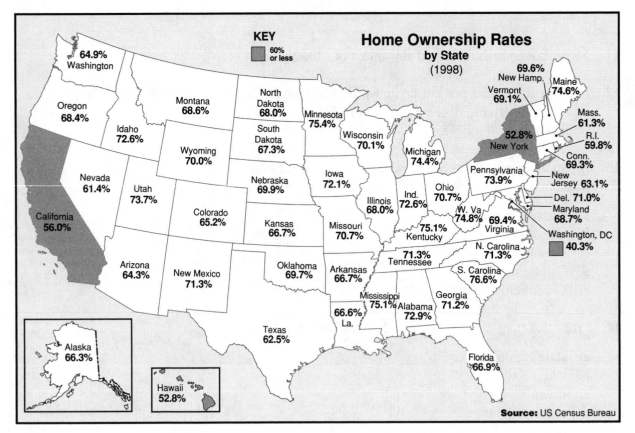

KEY
☐ 60% or less

Home Ownership Rates
by State
(1998)

Washington 64.9%
Oregon 68.4%
Montana 68.6%
North Dakota 68.0%
Minnesota 75.4%
New Hamp. 69.6%
Vermont 69.1%
Maine 74.6%
Idaho 72.6%
Wyoming 70.0%
South Dakota 67.3%
Wisconsin 70.1%
Michigan 74.4%
Mass. 61.3%
New York 52.8%
R.I. 59.8%
Conn. 69.3%
Nevada 61.4%
Utah 73.7%
Nebraska 69.9%
Iowa 72.1%
Illinois 68.0%
Ind. 72.6%
Ohio 70.7%
Pennsylvania 73.9%
New Jersey 63.1%
Del. 71.0%
California 56.0%
Colorado 65.2%
Kansas 66.7%
Missouri 70.7%
W. Va 74.8%
Virginia 69.4%
Maryland 68.7%
Kentucky 75.1%
Washington, DC ☐ 40.3%
Arizona 64.3%
New Mexico 71.3%
Oklahoma 69.7%
Arkansas 66.7%
Tennessee 71.3%
N. Carolina 71.3%
S. Carolina 76.6%
Mississippi 75.1%
Alabama 72.9%
Georgia 71.2%
La. 66.6%
Texas 62.5%
Florida 66.9%
Alaska 66.3%
Hawaii 52.8%

Source: US Census Bureau

Answer each question by filling in the blank, answering true or false, or choosing the best multiple-choice response.

17. The map above shows the percent of _____ for each state in the United States during the year _____.

18. States that are shaded had home ownership rates of _____.

19. In 1998 Florida's home ownership rate was 66.9%.   True   False

20. Idaho's 1998 home ownership rate was less than 70%.   True   False

21. Which state shown has the highest home ownership rate during 1998?

    a. South Carolina
    b. Minnesota
    c. West Virginia
    d. Texas
    e. Alaska

22. In which area is the home ownership rate 66% or greater?

    a. northwest states
    b. southwest states
    c. south central states
    d. southeast states
    e. northeast states

23. A family moving from Kentucky to New York is _____ to have an opportunity to own a home than a family moving from Hawaii to Mississippi.

    a. more likely
    b. less likely
    c. equally likely
    d. never likely
    e. 100% sure

24. During 1998, the home ownership rate in Alaska was _____ lower than the home ownership rate in Iowa.

    a. 4.8%
    b. 5.8%
    c. 3.5%
    d. 6.3%
    e. 6.0%

Map Review Chart

Circle the number of any problem that you missed and review the appropriate pages. A passing score is 20 correct answers. If you miss more than 4 questions, you should review the chapter.

PROBLEM NUMBERS	SKILL AREA	PRACTICE PAGES
1, 2, 3, 4, 5, 6, 7, 8	geographical map	106–123
9, 10, 11, 12, 13, 14, 15, 16	directional map	106–117, 124–129
17, 18, 19, 20, 21, 22, 23, 24	informational map	106–117, 130–135

Posttest A

This posttest will show you how well you have learned all of the skills that you have practiced in this book. Take your time and work each problem carefully. Answer the questions by filling in the blank, answering true or false, or choosing the best multiple-choice response.

GRAPH A

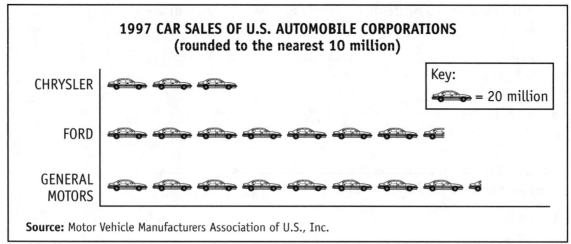

1997 CAR SALES OF U.S. AUTOMOBILE CORPORATIONS
(rounded to the nearest 10 million)

Key: = 20 million

Source: Motor Vehicle Manufacturers Association of U.S., Inc.

1. Graph A shows the passenger car production by _____, with the symbol 🚗 representing _____ cars.

2. In 1997 General Motors Corporation sold 150 million cars. True False

3. According to Graph A, for every Chrysler sold, at least _____ Fords were sold.

    a. 8
    b. 7
    c. 3
    d. 2
    e. 1

4. Which statement best describes Graph A?

    a. General Motors Corporation makes more money than any other company.
    b. Chrysler is projected to make the greatest gains in sales during the 2000s.
    c. General Motors sells twice as many cars per year as Chrysler.
    d. General Motors and Ford outsold Chrysler by a wide margin in 1997.

GRAPH B

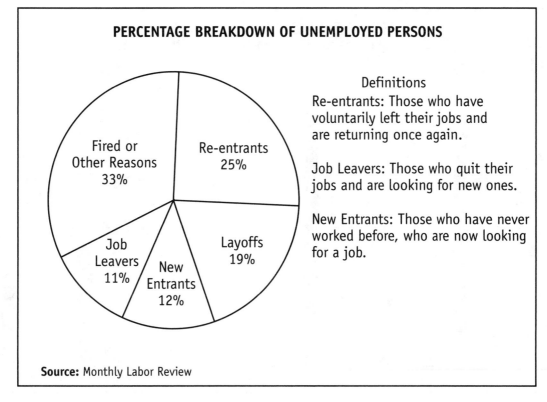

PERCENTAGE BREAKDOWN OF UNEMPLOYED PERSONS

Fired or
Other Reasons
33%

Re-entrants
25%

Job
Leavers
11%

New
Entrants
12%

Layoffs
19%

Definitions
Re-entrants: Those who have
voluntarily left their jobs and
are returning once again.

Job Leavers: Those who quit their
jobs and are looking for new ones.

New Entrants: Those who have never
worked before, who are now looking
for a job.

Source: Monthly Labor Review

5. Graph B shows the percentage breakdown of _____
   persons.

6. One-quarter of the unemployed are workers who are reentering   True   False
   the labor force, such as women who have taken time off to bear
   children.

7. The total percentage of unemployed who have been laid off is
   almost _____ than those who are leaving their jobs. (**Hint:** Round
   percentages to the nearest ten.)

   a. one-third less
   b. two times greater
   c. three times less
   d. three times greater
   e. one-half less

8. Which statement best describes Graph B?

   a. The largest group of the unemployed is made up of re-entrants.
   b. The smallest percentage of the unemployed is new entrants.
   c. The categories Layoffs and Fired or Other Reasons together
      make up the majority of the unemployed.
   d. The total percentage of those unemployed is composed of job
      leavers, new entrants, and re-entrants.
   e. More people are re-entering the job force than ever before.

GRAPH C

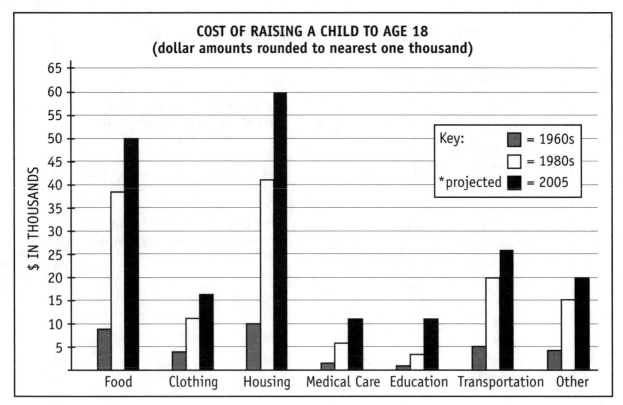

COST OF RAISING A CHILD TO AGE 18
(dollar amounts rounded to nearest one thousand)

9. Graph C shows the cost of raising a child for a total of
   _____ years.
   (number)

10. According to Graph C, the greatest increase between 1960   True   False
    and 2005 is in food costs.

11. Projecting to the year 2005 from the 1980s, what will be the
    increase in the cost of raising a child to age 18?

    a. $35,000   d. $158,000
    b. $40,000   e. $210,000
    c. $57,000

12. All the following statements can be concluded from Graph C
    EXCEPT:

    a. The cost of raising a child to age 18 is projected to increase at a
       more rapid rate than in the 1960s and 1980s.
    b. The cost of raising a child in the 1980s was more than
       three times as expensive as it was in the 1960s.
    c. Housing was the most costly area across all three time periods.
    d. The cost of raising a child will continue to increase at three
       times the rate as in the 1960s and 1980s.
    e. The cost of food was four times greater in the 1980s than in
       the 1960s.

GRAPH D

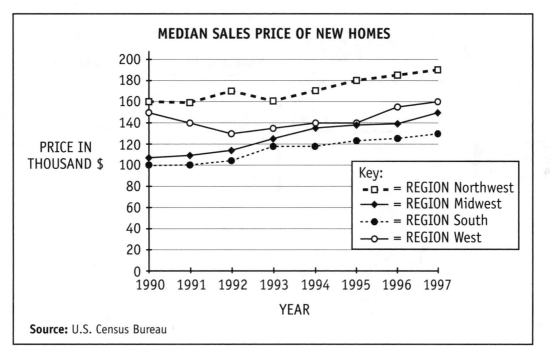

MEDIAN SALES PRICE OF NEW HOMES

PRICE IN THOUSAND $

Key:
- ▫ = REGION Northwest
- ◆ = REGION Midwest
- ● = REGION South
- ○ = REGION West

YEAR

Source: U.S. Census Bureau

13. Consistently, the most expensive homes shown on Graph D are in the _____.

14. For the years shown on the graph, 1994–1995 saw the smallest increase in price of home sales in the West.   True   False

15. In what year did the price of homes in the West come closest to the price of homes in the Midwest?

    a. 1996
    b. 1995
    c. 1994
    d. 1993
    e. 1992

16. Which statement is true according to Graph D?

    a. By 1995 the median price of homes across the United States was greater than $125,000.
    b. The gap between home prices in the Northwest and the South narrowed considerably between 1995 and 1997.
    c. House prices in general are much higher in the Midwest and the South than in other parts of the U.S.
    d. Following price trends, the median house price in the Midwest will probably be as much as the median house price in the Northwest by the year 2000.
    e. According to Graph D, it is probable that the median house price in the Northwest will reach $200,000 by the year 2005.

SCHEDULE A

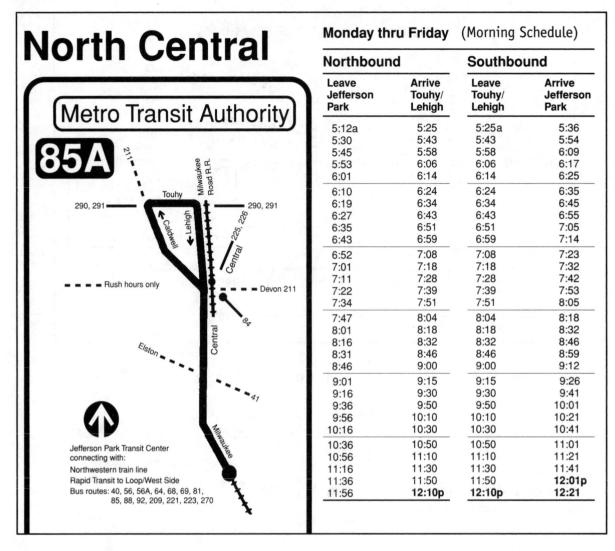

North Central

Metro Transit Authority

85A

211

Touhy

290, 291 — Milwaukee Road R.R.

290, 291

Lehigh

Caldwell

225, 226

Central

- - - Rush hours only

Devon 211

84

Central

Elston

41

Jefferson Park Transit Center
connecting with:
Northwestern train line
Rapid Transit to Loop/West Side
Bus routes: 40, 56, 56A, 64, 68, 69, 81,
  85, 88, 92, 209, 221, 223, 270

Monday thru Friday (Morning Schedule)

Northbound		Southbound	
Leave Jefferson Park	Arrive Touhy/ Lehigh	Leave Touhy/ Lehigh	Arrive Jefferson Park
5:12a	5:25	5:25a	5:36
5:30	5:43	5:43	5:54
5:45	5:58	5:58	6:09
5:53	6:06	6:06	6:17
6:01	6:14	6:14	6:25
6:10	6:24	6:24	6:35
6:19	6:34	6:34	6:45
6:27	6:43	6:43	6:55
6:35	6:51	6:51	7:05
6:43	6:59	6:59	7:14
6:52	7:08	7:08	7:23
7:01	7:18	7:18	7:32
7:11	7:28	7:28	7:42
7:22	7:39	7:39	7:53
7:34	7:51	7:51	8:05
7:47	8:04	8:04	8:18
8:01	8:18	8:18	8:32
8:16	8:32	8:32	8:46
8:31	8:46	8:46	8:59
8:46	9:00	9:00	9:12
9:01	9:15	9:15	9:26
9:16	9:30	9:30	9:41
9:36	9:50	9:50	10:01
9:56	10:10	10:10	10:21
10:16	10:30	10:30	10:41
10:36	10:50	10:50	11:01
10:56	11:10	11:10	11:21
11:16	11:30	11:30	11:41
11:36	11:50	11:50	**12:01p**
11:56	**12:10p**	**12:10p**	**12:21**

17. Schedule A shows the bus schedule for the route number _____,
    which is named _____.

18. The Northbound bus leaves from Touhy/Lehigh Streets.   True   False

19. According to Schedule A, the time it takes to go from Jefferson Park
    to Touhy/Lehigh is

    a. 13 minutes   **c.** 15 minutes   **e.** varied with the time of the morning
    b. 14 minutes   **d.** always the same

20. What is the total time it takes the bus leaving at 10:56 from
    Jefferson Park to make a round trip?

    a. 20 minutes   **d.** 25 minutes
    b. 24 minutes   **e.** 35 minutes
    c. 31 minutes

CHART A

A Nutritive Value of Dairy Foods					
Food Product	**Calories**	**Protein (g)**	**Fat (g)**	**Calcium (g)**	**Vitamin A**
Cheese-Cheddar (1 oz)	115	7	9	204	300 (I.U.)
Cottage Cheese (1 cup)	220	26	9	126	340
Swiss Cheese (1 oz)	105	8	8	272	240
Whole Milk (1 cup)	150	8	8	291	310
Nonfat Milk (1 cup)	85	8	T	302	500
Ice Cream (1 cup)	270	5	14	176	540
Sherbet (1 cup)	270	2	4	103	190

Source: U.S. Department of Agriculture, *Home and Garden Bulletin No. 72*

21. Chart A is based on information from the U.S. Dept. of
_____.

22. From Chart A, you can determine the number of calories in   True   False
one slice of Swiss cheese.

23. The difference in calcium between whole milk and nonfat milk is
_____ grams per cup.

   a. 11   **b.** 75   **c.** 83   **d.** 19   **e.** 146

24. Which graph below most accurately reflects the chart above?

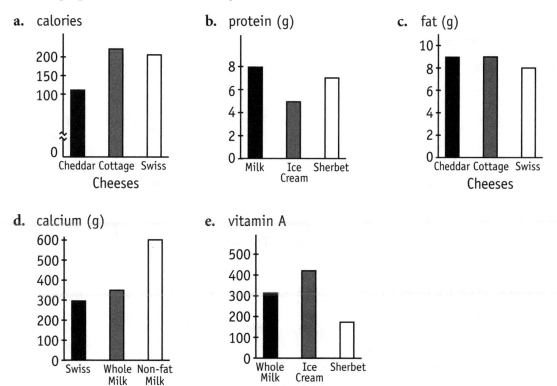

MAP A

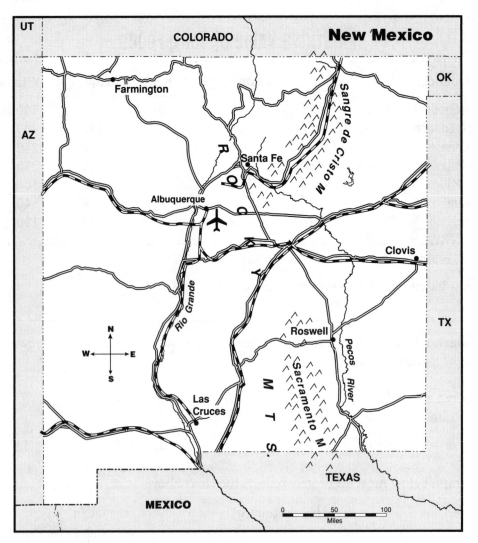

25. Map A shows the state of New Mexico, bordered by Utah, _____, Oklahoma, Texas, and _____.

26. There is an airport in the city of Albuquerque.   True   False

27. Where are the Sacramento Mountains located?

    a. east of the Rio Grande River   **c.** north of Albuquerque

    b. in the northeast part of the state   **d.** west of Las Cruces

28. It can be concluded from Map A that all of the following statements are true EXCEPT:

    a. The Rio Grande River crosses both the north and south borders of New Mexico.

    b. To travel from Santa Fe to Farmington, one must cross the Sangre de Cristo Mountains.

    c. The distance from Santa Fe to Albuquerque is less than 75 miles.

MAP B

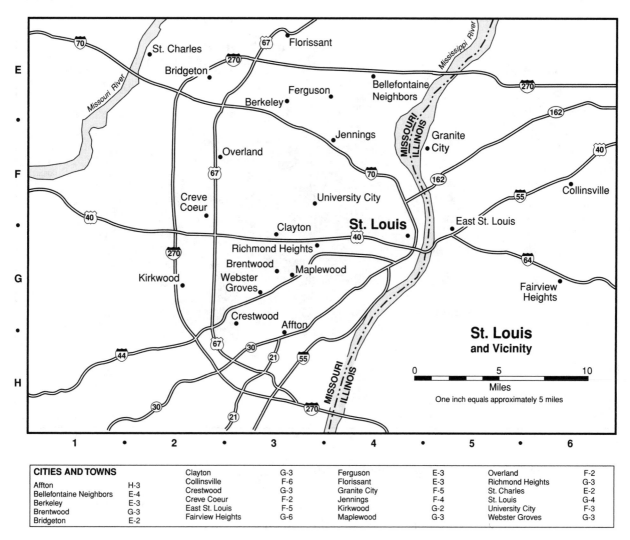

CITIES AND TOWNS

Affton	H-3	Clayton	G-3	Ferguson	E-3	Overland	F-2
Bellefontaine Neighbors	E-4	Collinsville	F-6	Florissant	E-3	Richmond Heights	G-3
Berkeley	E-3	Crestwood	G-3	Granite City	F-5	St. Charles	E-2
Brentwood	G-3	Creve Coeur	F-2	Jennings	F-4	St. Louis	G-4
Bridgeton	E-2	East St. Louis	F-5	Kirkwood	G-2	University City	F-3
		Fairview Heights	G-6	Maplewood	G-3	Webster Groves	G-3

29. Webster Groves is located in map section _____.

(letter-number)

30. Granite City is located in Illinois. True False

31. Richmond Heights is located

    a. 3 miles northeast of University City
    b. less than 20 miles west of St. Louis
    c. less than 20 miles east of St. Louis
    d. 6 miles south of Florissant
    e. on the border between Missouri and Illinois

32. Affton is located at the junction of ㉚ and what other highway?

    a. ㊵   **b.** ⑵⁷⁰   **c.** ㉑   **d.** ⑬   **e.** ⑮

MAP C

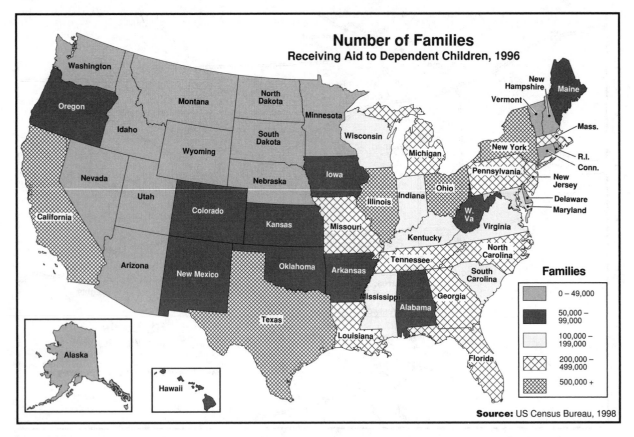

Number of Families
Receiving Aid to Dependent Children, 1996

Families

	0 – 49,000
	50,000 – 99,000
	100,000 – 199,000
	200,000 – 499,000
	500,000 +

Source: US Census Bureau, 1998

33. Map C shows the number of families receiving _____.

34. More than 100,000 but less than 500,000 families with dependent   True   False
    children receive federal funds in the state of California.

35. States in which 50,000 to 99,000 families receive assistance are
    primarily located in the _____ United States.

   a. western   d. central
   b. northern   e. southern
   c. eastern

36. From Map C you can conclude the following:

   a. States with fewer than 100,000 families receiving assistance tend
      to be in the northwestern states.
   b. Displaced farm families constitute the majority of people
      receiving aid.
   c. States with the greatest number of families receiving aid have
      the highest rates of unemployment.
   d. States with the largest populations tend to have the highest
      number of families receiving aid.
   e. Ohio has more families receiving aid than any other state.

Posttest A Prescriptions

Circle the number of any problem that you miss. A passing score is 30 correct answers. If you passed the test, go on to Using Number Power. If you did not pass the test, review the chapters in the book or refer to these practice pages in other materials from Contemporary Books.

PROBLEM NUMBERS	SKILL AREA	PRACTICE PAGES
1, 2, 3, 4, 5, 6, 7, 8, 9	graphs	16–67
10, 11, 12, 13, 14, 15, 16	Real Numbers: Tables, Graphs, and Data Interpretation	12–43
	Math Exercises: Data Analysis and Probability	10–19
	Math Skills That Work: Book Two	160–179
17, 18, 19, 20, 21, 22, 23, 24	schedules and charts	80–95
	Real Numbers: Tables, Graphs, and Data Interpretation	1–11
	Math Exercises: Data Analysis and Probability	3–9
25, 26, 27, 28, 29, 30, 31, 32, 33, 34, 35, 36	maps	106–135

For further graph, chart, schedule, and map practice:

Math Solutions (software)
Graphic Data

Pre-GED/Basic Skills Interactive (software)
Mathematics Unit 8
Social Studies Unit 5

GED Interactive (software)
Reading Unit 4
Social Studies Unit 1

Posttest B

This review test has a multiple-choice format much like the GED and other standardized tests. Take your time and work each problem carefully. Circle the correct answer to each problem. When you finish, check your answers at the back of the book.

GRAPH A

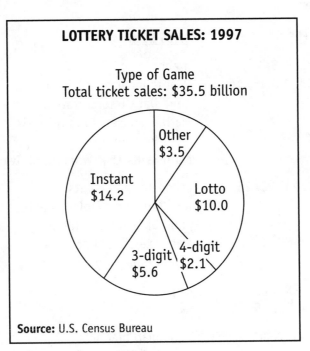

LOTTERY TICKET SALES: 1997

Type of Game
Total ticket sales: $35.5 billion

Other $3.5
Instant $14.2
Lotto $10.0
3-digit $5.6
4-digit $2.1

Source: U.S. Census Bureau

1. Graph A shows total lottery ticket sales by

   a. U.S. Census Bureau
   b. percent of money spent per household
   c. number of tickets sold
   d. type of game
   e. 1990 through 1997

2. Total ticket sales for 1997 was

   a. $14.2 billion   b. $35.5 billion   c. $21.1 billion   d. $29.9 billion   e. $10 billion

3. Which type of lottery game accounts for more than $\frac{1}{3}$ of total sales?

   a. 3-digit   b. 4-digit   c. instant   d. lotto   e. other

4. Which statement best describes the contents of Graph A?

   a. Instant lottery tickets in 1997 cost $14.20.
   b. $3.5 billion was spent on things other than lottery tickets in 1997.
   c. Lottery ticket sales in 1997 totaled $35.5 billion.
   d. Lottery ticket sales increases every year.
   e. Lotto and 4-digit lottery games were more popular than any other games.

GRAPH B

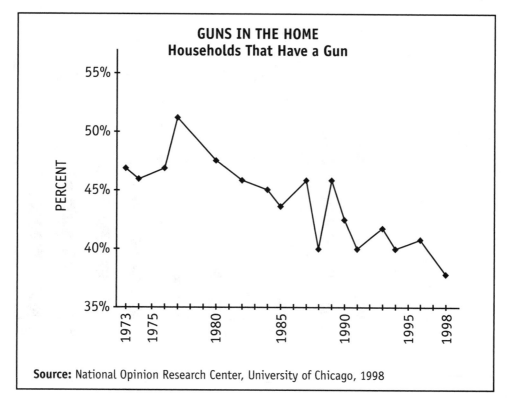

GUNS IN THE HOME
Households That Have a Gun

PERCENT

Source: National Opinion Research Center, University of Chicago, 1998

5. Graph B shows a steady decrease in

   a. number of people who own guns in 1998
   b. percent of guns found in homes
   c. number of guns used in crimes from 1973 to 1998
   d. percent of household guns used in crimes
   e. percent of households that have a gun

6. The percent of households with guns dropped _____ percent from 1977 to 1998.

   a. 25   **b.** 32   **c.** 14   **d.** 5   **e.** 52

7. If the percent of guns continues to decline at the same rate as in the period between 1990 to 1998, one can expect the percentage in 2006 to be

   a. 38–40%   **c.** 10–20%   **e.** 0–5%
   b. 30–35%   **d.** 5–10%

8. Which statement is supported by the contents of Graph B?

   a. More and more people support a ban on handguns.
   b. More households own guns now than in the 1930s.
   c. About 50% of all American households owned guns in the 1980s.
   d. The number of households with a gun has declined since 1977.
   e. People increasingly support the right to bear arms.

GRAPH C

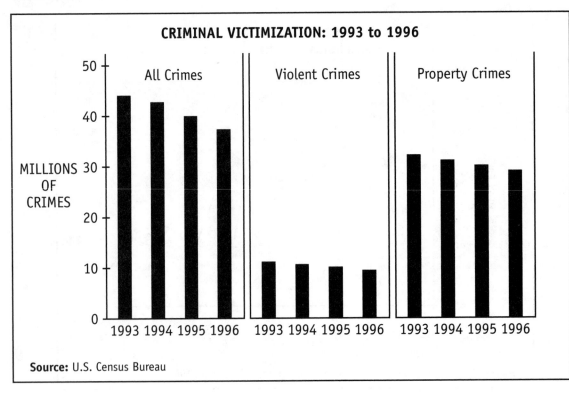

CRIMINAL VICTIMIZATION: 1993 to 1996

All Crimes | Violent Crimes | Property Crimes

MILLIONS OF CRIMES

Source: U.S. Census Bureau

9. Graph C shows the number of _____ and _____ crimes for a 4-year period.

   a. private; total
   b. violent; property
   c. violent; rape
   d. burglary; violent
   e. gun; property

10. Total crimes during this 4-year period have dropped by approximately

   a. 30 million
   b. 20 million
   c. 5 million
   d. 10 million
   e. 15 million

11. In 1996 violent crimes represented approximately _____ percent of all crime. (**Hint:** Round to nearest percent.)

   a. 25   **b.** 75   **c.** 50   **d.** 40   **e.** 33

12. All of the following statements can be concluded for Graph C EXCEPT:

   a. Crime in all categories show a consistent decrease.
   b. There are more crimes committed against property than violent crimes.
   c. Of the years shown, 1996 had the least total crimes.
   d. Crime has decreased by more than 10% during this 4-year period.
   e. Crime will continue to decrease in the late 1990s.

GRAPH D

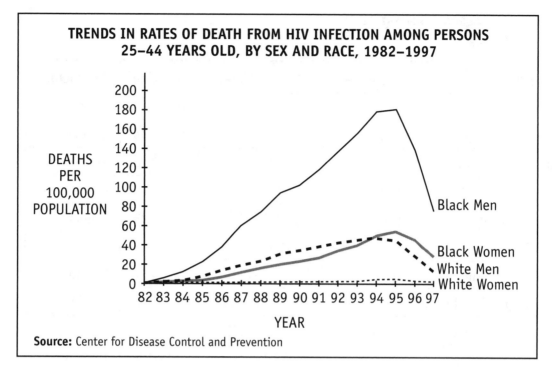

TRENDS IN RATES OF DEATH FROM HIV INFECTION AMONG PERSONS 25–44 YEARS OLD, BY SEX AND RACE, 1982–1997

DEATHS PER 100,000 POPULATION

Black Men
Black Women
White Men
White Women

YEAR

Source: Center for Disease Control and Prevention

13. Graph D shows the number of deaths per _____ population of individuals infected with HIV (AIDS).

 a. 25–44   **c.** 100,000   **e.** 20 million
 b. one million   **d.** 200

14. The trend in HIV death rates for black men grew _____ and then recently _____.

 a. slightly; increased   **c.** rapidly; increased   **e.** drastically; fell slightly
 b. rapidly; declined   **d.** slightly; stayed the same

15. Between the years 1995 and 1996, black men showed about _____ more HIV-related deaths than black women.

 a. 2 times   **c.** 5 times   **e.** 3 times
 b. one-half   **d.** one-third

16. Which statement is true according to Graph D?

 a. U.S. citizens show more HIV-related deaths than other groups.
 b. On the average, for every 180 deaths that occur in 100,000 black men, 10 to 15 deaths occur in 100,000 white men.
 c. In persons 25–44 years old, the trend in HIV-related deaths has shown major decreases recently.
 d. More HIV-related deaths occur in the U.S. than in other countries.
 e. No HIV-related deaths have occurred among persons over 44.

SCHEDULE A

Schedule of Classes—Spring 1999
Sociology (SOC)

Course Title	Course #	Meeting Time‡	Location	Instructor
Internship/Practicum	410	TBA*	TBA*	Staff
Sociological Theory	413	TR 1100–1200	MLM 234	Conroy
Conducting Social Research	416	TR 1230–1350	MCC 201	Kenny
Rural-Urban Sociology	475	T 1800–2050	STAG 109	Kramer
Law and Society	491	TR 1230–1350	MLM 123	Sellers
Social Projects	506	TBA*	TBA*	Staff
Leisure and Culture	513	TR 1100–1220	MLM 234	Pershing
Conducting Social Research	516	TR 1230–1350	MCC 201	Plankford
Juvenile Delinquency	540	TR 1400–1520	STAG 111	Conway
Sociology of Religion	552	MWF 1300–1350	FAIR 305	Langley
Leisure and Culture	554	M 1900–2150	STAG 109	Mitchell
Gender Issues	575	T 1800–2050	STAG 109	Crailer

*TBA = to be announced
‡using a 24-hr clock, 1:30 P.M. would be 1330; M = Monday, T = Tuesday, W = Wednesday, R = Thursday, F = Friday

17. Schedule A shows the range of courses that can be studied within what subject area?

    a. Spanish   **b.** Forestry   **c.** Law   **d.** Sociology   **e.** Religion

18. To register for the course called Juvenile Delinquency, one must register for what course number?

    a. Conway   **b.** 540   **c.** STAG 111   **d.** 410   **e.** 1400–1520

19. According to Schedule A, _____ and _____ are two courses whose meeting times are to be announced.

    a. SOC 410 and SOC 516   **d.** SOC 410 and SOC 506
    b. SOC 506 and SOC 552   **e.** SOC 416 and SOC 516
    c. SOC 410 and SOC 575

20. According to Schedule A, which course meets in the evening hours on Monday?

    a. SOC 554   **b.** SOC 513   **c.** SOC 475   **d.** SOC 575   **e.** SOC 491

CHART A

TRAFFIC: IT'S GOING TO GET WORSE

The increase in average daily traffic on selected roadways, according to TransPlan projections.

Roadway	1995	2015	% Increase
1. West Sixth-Seventh (at Chambers)	66,000	84,200	28
2. Belt Line (east of Delta)	56,600	81,000	43
3. Interstate 5 (south of Belt Line)	49,800	71,500	44
4. Ferry Street Bridge	60,300	75,000	24
5. Interstate 105 (at Pioneer Parkway)	46,000	67,500	47
6. Belt Line (west of River Road)	34,400	53,700	56
7. West 11th Ave. (at Seneca)	38,200	48,000	26
8. Springfield-Glenwood bridge	27,500	44,900	63
9. Coburg Road (at Harlow)	35,100	41,200	17
10. Gateway Boulevard	24,000	38,000	58
11. Barger Drive (west of Belt Line)	19,800	25,700	30
12. Pioneer Parkway	12,800	23,600	84
13. Chad Drive (at Costco)	8,400	15,900	89
14. Hayden Bridge Road (at Fifth St.)	9,000	14,400	60
15. Green Hill Road (at Barger)	4,600	13,200	187
16. Delta Highway (north of Belt Line)	5,200	12,500	140
17. Willow Creek Road (at W. 11th)	1,400	11,500	721
18. Jasper Road	7,400	11,200	51
19. Coburg Road (at Armitage Park)	6,400	10,000	56

Source: Lane Council of Governments

21. Chart A shows the average daily _____ on selected roadways.

   a. patterns　　**b.** destinations　**c.** traffic　　**d.** buses　　**e.** drivers

22. Two roadways projected to more than double in daily traffic by 2015 are

   a. Willow Creek Road and Barger Drive
   b. Interstate 105 and Green Hill Road
   c. Belt Line (east of Delta) and Springfield-Glenwood bridge
   d. Delta Highway and West 11th Avenue
   e. Willow Creek Road and Green Hill Road

23. The area to move into with the least projected increase in traffic by 2015 would be

   a. West 11th Avenue at Seneca
   b. Coburg Road at Armitage Park
   c. Belt Line west of River Road
   d. Chad Drive at Costco
   e. Willow Creek Road at West 11th

MAP A

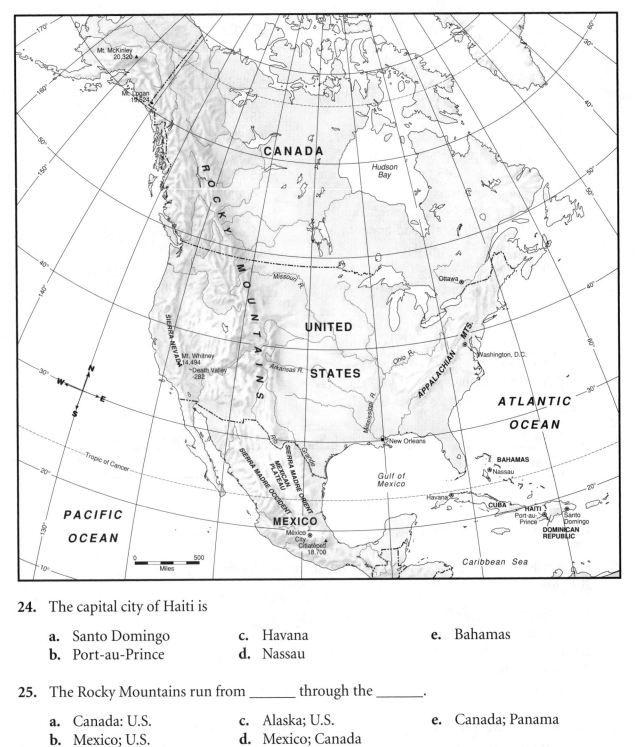

24. The capital city of Haiti is

    a. Santo Domingo
    b. Port-au-Prince
    c. Havana
    d. Nassau
    e. Bahamas

25. The Rocky Mountains run from _____ through the _____.

    a. Canada: U.S.
    b. Mexico; U.S.
    c. Alaska; U.S.
    d. Mexico; Canada
    e. Canada; Panama

26. All the statements below are true EXCEPT:

    a. Mexico City is 1,500–2,000 miles from Washington, D.C.
    b. All countries in North America are mountainous.
    c. A river acts as a boundary between Mexico and the U.S.
    d. Mt. Whitney is the highest mountain in North America.

MAP B

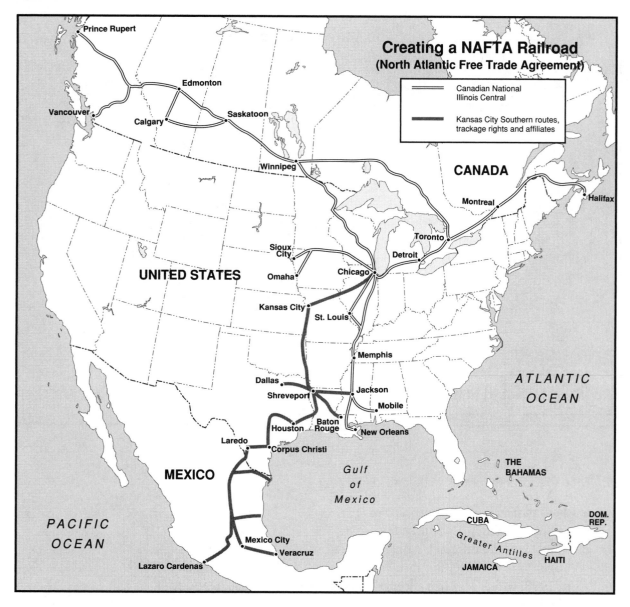

Creating a NAFTA Railroad
(North Atlantic Free Trade Agreement)

27. Map B shows the NAFTA Railroad extending to cities as far north
    as _____, Canada, and as far south as _____, Mexico.

    a. Toronto; Mexico City
    b. Prince Rubert; Veracruz

    c. St. Louis; Detroit
    d. Vancouver; Halifax

28. All the following statements below are true EXCEPT:

    a. The proposed NAFTA Railroad will be operated by the
       Canadian National Illinois Central and Kansas City Southern.
    b. The proposed NAFTA Railroad travels through three countries.
    c. The proposed NAFTA Railroad concentrates travel in east and
       west Canada, in central U.S., and in east Mexico.
    d. The proposed NAFTA Railroad travels through the largest cities
       of Canada, the U.S., and Mexico.

MAP C

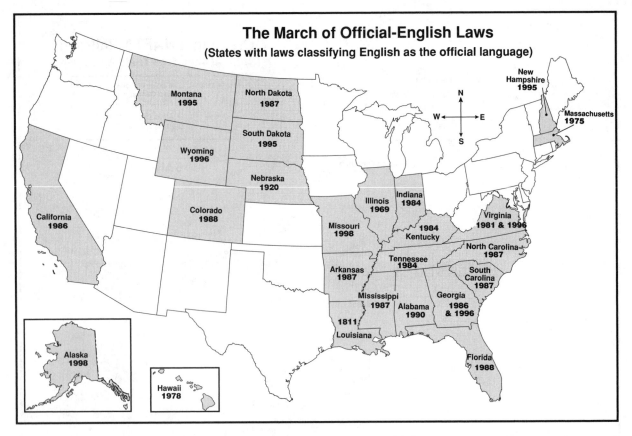

The March of Official-English Laws
(States with laws classifying English as the official language)

29. Map C shows the states with laws classifying

    a. laws passed in each state
    b. years when they became states
    c. English as the official language

30. As of 1998, which area has adopted laws classifying English as the
    official language?

    a. all of the midwestern states
    b. all of the southeastern states
    c. all of the southeastern and midwestern states

    d. all of the northeastern states
    e. all of the southern states

31. The first and the most recent states that adopted laws of English as
    their official language are

    a. Missouri and Alaska
    b. Louisiana and Virginia
    c. Alaska and Mississippi

    d. Nebraska and Louisiana
    e. Louisiana and Alaska

32. Which graph best describes Map C?

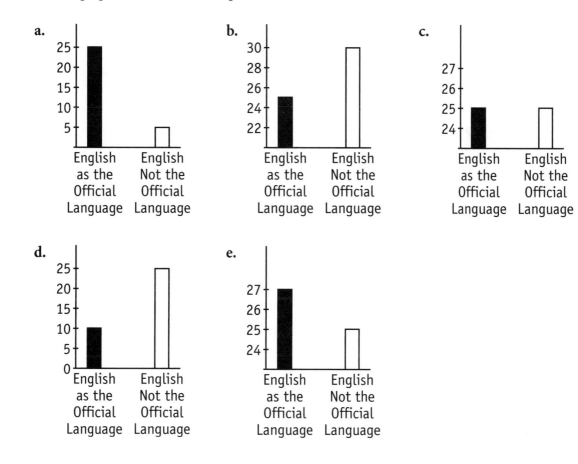

Using
Number
Power

Budgeting a Paycheck

Budgeting money is an important financial skill. Successful individuals and companies use budgeting to plan the flow of money.

Rather than spending money in an unplanned way, budgeting allows an individual to carefully plan the percent of income that should go for food, housing, clothing, and other important monthly costs. The circle graph below is an example of family budgeting.

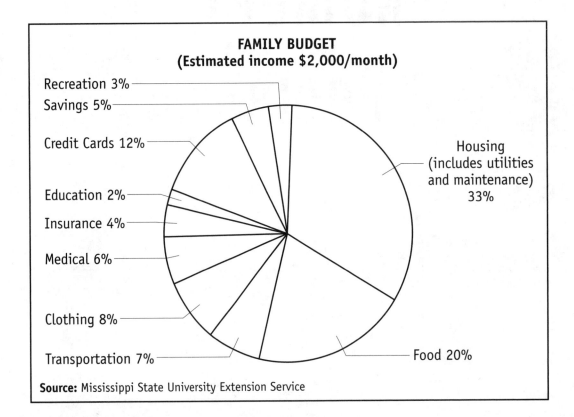

FAMILY BUDGET
(Estimated income $2,000/month)

Recreation 3%
Savings 5%
Credit Cards 12%
Education 2%
Insurance 4%
Medical 6%
Clothing 8%
Transportation 7%

Housing (includes utilities and maintenance) 33%

Food 20%

Source: Mississippi State University Extension Service

For questions 1–4, answer true or false or choose the best multiple-choice response.

1. According to the graph, the percentages for housing and food costs add up to slightly more than half of the family's total expenditures.

   True   False

2. After housing and food, medical care is the largest budgeted family expenditure.

   True   False

3. If a family's take-home pay was $2,000, the budget allowed the family to spend up to _____ for food.

   a. $200   b. $300   c. $210   d. $100   e. $400

4. This family did not have any medical expenses this month. They decided to put the money budgeted for medical expenses into savings. What percent of the family's income went into savings for the month?

   a. 21.5%   b. 15.5%   c. 25.5%   d. 11%   e. 5%

5. Last month, the Smith family made the following expenditures and spent over their income of $2,000. Using the information on the graph, circle the items that were incorrectly spent and make the necessary corrections for the following month's new budget. Remember, a "balanced" budget means total expenses equal total income.

INCOME: TAKE-HOME PAY + $2,000			
Expenses		Adjustments (if any)	Corrected Budget
Food	$400	0	$400
Clothing	$170	−$10	$160
Insurance	$ 80		
Housing	$660		
Medical	$140		
Transportation	$145		
Recreation	$200		
Savings	$100		
Credit Cards	$240		
Education	$ 40		
Total Expenses	$2,175		

Cost of Housing: Buying a House

Rising housing costs have made it impossible for many people to buy a home. Housing costs have been affected by a number of factors, including high interest rates for loans and increased costs of building materials.

The cost of housing varies widely across the country. It is generally recommended that no more than 25–30% of a family's income should go for housing. However, in some cities, the percentage of income that goes for housing greatly exceeds the recommended maximum percentage. As a result, many families must do without other things in order to make house payments.

		HOUSING COSTS, INCOME, AND MORTGAGE PAYMENTS		
Cities	Median Sales Price of One Family Housing*	Average Family Income per Year**	Median Monthly House Payment	Percentage of Income on House Payments
San Francisco	$292,600	$63,000	$1,680	32%
Washington, D.C.	$166,300	$50,250	$1,005	24%
Denver	$140,000	$45,780	$763	20%
Minneapolis	$118,400	$53,810	$852	19%
Chicago	$158,900	$48,380	$887	22%
Atlanta	$108,400	$53,530	$803	18%
Seattle	$171,300	$51,258	$897	21%

*National Association of Realtors   **Assuming two incomes per family

Source: U.S. Census Bureau

 For questions 1–4, answer each question by filling in the blank or choosing the best multiple-choice response.

1. The median sale price of a home in San Francisco is
   $_____ more per year than for a family in Seattle.

2. Basing your answers on the chart, who has more money to spend after the house payment is made: a family living in Denver or a family living in Atlanta? (**Hint:** First find monthly income by dividing yearly income by the number of months per year.)

3. Based on the chart, the percentage of income paid for house payments in Minneapolis is _____ than in San Francisco.

   a. 22.0% more   b. 15% less   c. 31.9% more   d. 13% less   e. 27.5% less

4. The following conclusion can be drawn from the chart:

   a. The percentage of income spent on house payments varies regionally across the United States.
   b. Average family incomes vary according to the tax burden assumed by the family.
   c. In the seven cities listed, the percentage of income spent on housing exceeds the recommended maximum, creating a financial burden on families.
   d. The percentage of two-income families is increasing rapidly.
   e. The cost of existing housing is expected to decline.

5. Based on the chart below, decide whether you could afford the monthly house payments along with other family expenses, if you were

Living in	and Your Monthly* Income is	and Your House Payment is	+ Other Expenses	Answer	
Washington, D.C.	$3,310	$1,005	+ $2,032?	Yes	No
Chicago	$2,629	$887	+ $1,820?	Yes	No
Seattle	$2,482	$897	+ $1,635?	Yes	No
San Francisco	$2,622	$1,680	+ $1,400?	Yes	No
Minneapolis	$2,803	$852	+ $1,701?	Yes	No

*assuming two incomes per family

Cost of Utilities: Paying Your Electric Bill

One of the major concerns of people throughout the world is the cost of energy. Oil, gas, and electric rates have risen to record highs during the past several years.

Study the information below to determine how electric rates compare from one location in the country to another.

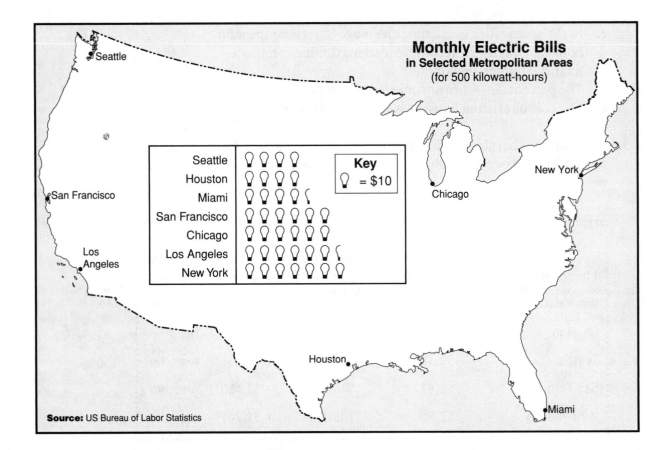

Monthly Electric Bills
in Selected Metropolitan Areas
(for 500 kilowatt-hours)

Key
💡 = $10

Source: US Bureau of Labor Statistics

Answer each question by filling in the blank, completing the chart, or choosing the best multiple-choice response.

1. If you were in New York City, the cost of your electric bill would be $_____ more per month than if you were in Houston.

2. Compute the following electric bills based on the number of kilowatt-hours. Remember the chart is based on a use of 500 kilowatt hours.

City	Kilowatt-Hours Used	Monthly Electric Bill
Houston	750 hours	$
Seattle	1,500 hours	$
Chicago	1,500 hours	$
Miami	2,000 hours	$
New York City	2,000 hours	$

3. If electric rates increase by 10%, what would be the total amount per month for 500-kilowatt-hour electric bills in the following cities?

Seattle $_____
San Francisco $_____
Los Angeles $_____

4. From the graph on page 168, you could conclude the following:

a. There are more electric power sources in the West, thus accounting for cheaper costs.
b. More people use electricity in the East than in the West.
c. Southern and northern states show little difference in electric power rates.
d. The cost of utilities in New York is greater than that in almost all other cities in the United States.
e. Electric bills tend to be higher in western cities than in eastern cities.

Cost of Utilities: Paying a Telephone Bill

One of the major costs in a budget is the telephone bill. While local calls can mount up, long distance calls can ruin your budget.

There are several guidelines you can follow to reduce your bill. Call at hours when there are reduced rates. This time is generally in the evenings. Also, if possible, make use of weekend rates, which are lower than on weekdays. Try to call station to station rather than calling person to person.

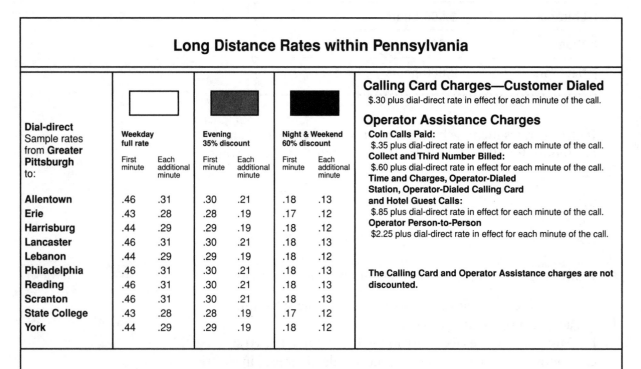

Long Distance Rates within Pennsylvania

Dial-direct Sample rates from **Greater Pittsburgh** to:	Weekday full rate		Evening 35% discount		Night & Weekend 60% discount	
	First minute	Each additional minute	First minute	Each additional minute	First minute	Each additional minute
Allentown	.46	.31	.30	.21	.18	.13
Erie	.43	.28	.28	.19	.17	.12
Harrisburg	.44	.29	.29	.19	.18	.12
Lancaster	.46	.31	.30	.21	.18	.13
Lebanon	.44	.29	.29	.19	.18	.12
Philadelphia	.46	.31	.30	.21	.18	.13
Reading	.46	.31	.30	.21	.18	.13
Scranton	.46	.31	.30	.21	.18	.13
State College	.43	.28	.28	.19	.17	.12
York	.44	.29	.29	.19	.18	.12

Calling Card Charges—Customer Dialed
$.30 plus dial-direct rate in effect for each minute of the call.

Operator Assistance Charges
Coin Calls Paid:
$.35 plus dial-direct rate in effect for each minute of the call.
Collect and Third Number Billed:
$.60 plus dial-direct rate in effect for each minute of the call.
Time and Charges, Operator-Dialed Station, Operator-Dialed Calling Card and Hotel Guest Calls:
$.85 plus dial-direct rate in effect for each minute of the call.
Operator Person-to-Person
$2.25 plus dial-direct rate in effect for each minute of the call.

The Calling Card and Operator Assistance charges are not discounted.

Rates to other states

Dial-direct Sample rates from **Greater Pittsburgh** to:	Weekday full rate		Evening 40% discount		Night & Weekend 60% discount	
	First minute	Each additional minute	First minute	Each additional minute	First minute	Each additional minute
Atlanta, GA	.62	.43	.37	.26	.24	.18
Atlantic City, NJ	.59	.42	.35	.26	.23	.17
Boston, MA	.62	.43	.37	.26	.24	.18
Chicago, IL	.59	.42	.35	.26	.23	.17
Des Moines, IA	.62	.43	.37	.26	.24	.18
Detroit, MI	.58	.39	.34	.24	.23	.16
Houston, TX	.64	.44	.38	.27	.25	.18
Los Angeles, CA	.74	.49	.44	.30	.29	.20
New York, NY	.59	.42	.35	.26	.23	.17
St. Louis, MO	.62	.43	.37	.26	.24	.18
Seattle, WA	.74	.49	.44	.30	.29	.20
Washington, DC	.58	.39	.34	.24	.23	.16

Calling Card Charges—Customer Dialed

Miles	Rate	Plus
1–10	$ .60	Dial direct mileage rate
11–22	.80	in effect for each minute
23 and beyond	1.05	of the call.

Operator Assistance Charges
Operator Station Third number billed, coin telephone, collect, time and charges, hotel guest calls and Operator-dialed Calling Card Calls.

Miles	Rate	Plus
1–10	$ .75	Dial direct mileage rate
11–22	1.10	in effect for each minute
23 and beyond	1.55	of the call.

Operator Person-to-Person

Miles	Rate	Plus
All	$3.00	Dial direct mileage rate in effect for each minute of the call.

The Calling Card and Operator Assistance charges are not discounted.

Use the long distance rate chart to answer the following questions.

1. You made 20 calls within the state of Pennsylvania last month, totaling $10. These calls were made on a weekday.
   How much money would you have saved if the same calls were made evenings or on weekends?

   evenings $_____   weekends: $_____

2. Compute the costs of the following monthly telephone calls based on the rate schedule on the previous page.

Mr. Bill Telephone Pittsburgh, Pennsylvania	Account Number: 123-456			Date Payment Due: January 9, 1999	
CURRENT MONTH'S CHARGES (W = weekday; E = evening; NW = night and weekend)					
In State Calls (direct dial)	**Code**	**Total Time**	**Amount Charged**	**Totals**	
Harrisburg	W	1 minute	$		
Erie	E	2 minutes	$		
York	E	3 minutes	$		
Allentown	W	2 minutes	$		
Harrisburg	W	1 minute	$		
			Total (In State Calls)	**$**	
Outside the State (direct dial)					
Los Angeles	NW	3 minutes	$		
Houston	W	1 minute	$		
			Total (Outside the State)	**$**	
Operator Assisted (person to person)					
Scranton	W	3 minutes	$		
Seattle	W	3 minutes	$		
			Total (Operator Assisted)	**$**	
Local Telephone Service:		From 11/14 to 12/14		**$**	15.15
			TOTAL NOW DUE	**$**	

Cost of Food

One factor that has caused the rate of inflation to rise is the increase in food prices. Although it is less expensive to purchase food to prepare at home than eating out, the price of basic foods has fluctuated over the last several years.

Most families with young children build their meals to include the basic food groups of dairy, protein, and vegetables. Ground beef and milk products are among the most often purchased foods for budget conscious families. The cost of these items, along with tomatoes, which are frequently used in sauces and stews, is tracked in the chart below.

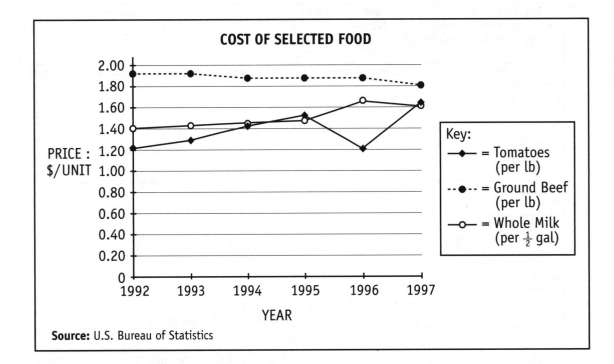

COST OF SELECTED FOOD

Key:
- —◆— = Tomatoes (per lb)
- --●-- = Ground Beef (per lb)
- —○— = Whole Milk (per ½ gal)

Source: U.S. Bureau of Statistics

Use the graph above to answer the following questions.

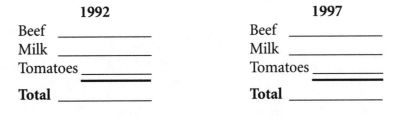

1. Your family eats 5 pounds of hamburger, 2 pounds of tomatoes, and 2 gallons of milk per week. Calculate the cost of these three food items for one week at 1992 prices and at 1997 prices.

	1992			**1997**
Beef	_____		Beef	_____
Milk	_____		Milk	_____
Tomatoes	_____		Tomatoes	_____
Total	_____		**Total**	_____

2. From question 1, find the difference in total costs for the selected food items between one week in 1992 and one in 1997.

3. From the graph, which two food items were about the same price between 1994 and 1995?

4. Compute the following food costs for 1 month if there are 4 weeks in the month and if your family consumes $4\frac{1}{2}$ pounds of beef, 2 gallons of milk, and 1 pound of tomatoes for dinner each week.

The Cost of Five Selected Foods for 1 Month: 1997

Beef	_____
Tomatoes	_____
Milk	_____
Potatoes	$5.00
Bread	$8.00
Total	_____

Costs of Operating a Car

Most people now realize that the money they pay to buy a car is only the beginning of their expenses. A buyer must also consider the cost of the daily wear and tear on the car, bills for maintaining and repairing the car, and the car's depreciation in value. These are factors in addition to monthly payments and the price of gas.

Depreciation is not as easily understood as a bill for repair or even the effect of wear and tear on a car. Depreciation is a term used to describe the decrease in dollar value of a car due to daily use. As a car gets older, it depreciates (decreases) in value.

The graph below shows the relative costs of operating a car. It also indicates that the number of miles that a car is driven per year tends to decrease over time. Why is this? One explanation is that today's cost-conscious drivers tend to "go easy" on an older car to guard against possible breakdowns.

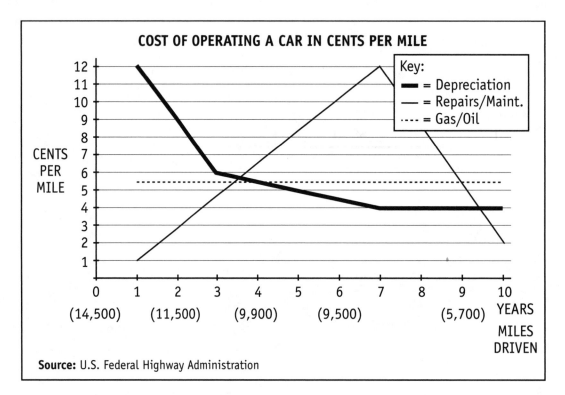

COST OF OPERATING A CAR IN CENTS PER MILE

Key:
━ = Depreciation
─ = Repairs/Maint.
···· = Gas/Oil

Source: U.S. Federal Highway Administration

Use the graph to answer the following questions.

1. Compute the cost of depreciation, repairs and maintenance, gas, and oil in cents per mile for the first year of ownership.

   Depreciation   _____ cents
   Repairs/Maintenance   _____ cents
   Gas/Oil   _____ cents

   Total   _____ cents per mile for the first year

2. If you drove the car 10,000 miles during the first year, what would be the cost of operating the car? (Base your answer on question 1.)
   $_____

3. During the _____ years, the cost of operating a car for the three categories is approximately the same.

   a. third and fourth
   b. fifth and sixth
   c. first and second
   d. sixth and seventh
   e. ninth and tenth

4. Determine the cost of operating your car during the seventh year.

Depreciation	_____ cents
Repairs/Maintenance	_____ cents
Gas/Oil	_____ cents
Total	_____ cents per mile for year 7

5. Compute the differences in operating costs between the first year (question 1) and the seventh year (question 4) in cents per mile.

   a. Which year is the most costly? (Circle one.)   Year 1   Year 7

   b. How much more costly? $_____ per mile.

   c. Based on your judgment and the graph, answer true or false.
      i. It is wise to sell a car after its seventh year to avoid   True   False
         major repair bills.
      ii. You would need to keep a car for 7 or more years to   True   False
         see total operating costs begin to drop.

Costs of Hospitalization

The cost of medical care is expensive and on the rise. Often the only protection against financial disaster is the purchase of a medical insurance policy. At first, monthly payments may seem an unnecessary cost, but the first major medical bill you face will seem much more manageable if you know that you will have help in paying it.

One of the main medical expenses you may face is the cost of hospitalization. The graph below compares the differences in hospital costs based on the kind of hospital surveyed.

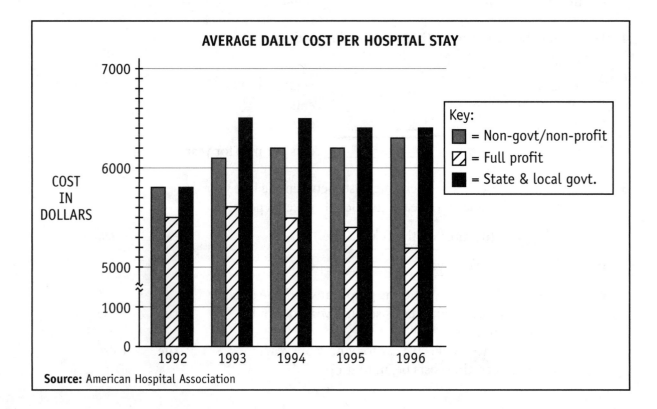

AVERAGE DAILY COST PER HOSPITAL STAY

Key:
■ = Non-govt/non-profit
▨ = Full profit
■ = State & local govt.

Source: American Hospital Association

Answer each question by filling in the blank.

1. Determine the cost of a 10-day stay at each of the three hospitals below. Use 1996 figures.

    Hospital A: Non-Govt.

    $_____

    Hospital B: Full Profit

    $_____

    Hospital C: State & Local Govt.

    $_____

2. Suppose in 1995 you had knee surgery. The hospital you were in was a full profit hospital. Your stay was 8 days.

What was the total expense for this hospitalization?

$_____

Your health insurance policy paid $34,560. How much did you owe after the insurance paid its portion of the bill?

$_____

3. Find the difference between 1992 and 1996 hospital costs in a non-profit hospital.

$_____ per day

Answer true or false to the following statements, basing your answer on the graph.

4. **a.** Non-government/non-profit hospitals have been consistently   True   False
   more expensive than the other two categories of hospitals in
   every year between 1992 and 1996.

   b. No category of hospital increased costs more than $200 per   True   False
   day over a one year period.

   c. The year in which all categories of hospitals show an increase   True   False
   in costs over the previous year was 1993.

Calculating Driving Distances and Times

Certain maps are designed to help the traveler estimate the distance in miles as well as the approximate time necessary to reach a specific place. On the map below, the numbers indicate the mileage between two cities. For example, the distance between Los Angeles and Phoenix is 389 miles.

Often the time to drive distance is calculated by estimating the total number of miles it takes to travel in one hour. For example, if you travel 50 miles in 1 hour, then it will take you 8 hours to travel a total of 400 miles. (50 miles × 8 hours = 400 miles)

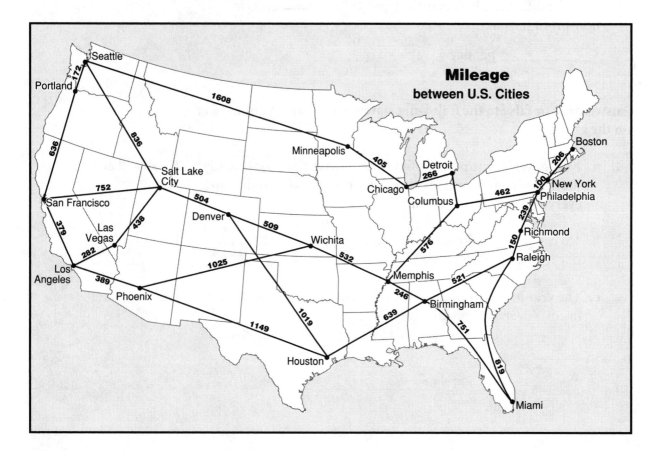

Answer each question by filling in the blank or choosing the best multiple-choice response.

1. Calculate the driving distances on the following trip. (Although many routes are possible, use the most direct route.)

From	To	Distance in Miles
Seattle	San Francisco	_____
San Francisco	Los Angeles	_____
Los Angeles	Houston	_____
	Total Miles	_____

2. Calculate the driving times for the following trips by dividing distance by miles per hour. (Round your answer to the nearest hour.)

From	To	Distance in miles	Time in hours (speed = 50 miles per hour)
Seattle	Minneapolis	_____	_____
San Francisco	Denver	_____	_____
Houston	Raleigh	_____	_____
Boston	Miami	_____	_____

3. You decide to take a trip from Phoenix to Wichita.

What is your gas mileage (in miles per gallon) if you used 50 gallons of gas (to the nearest tenth of a gallon)?

Note: $\dfrac{\text{total miles}}{\text{gallons used}} = \text{miles per gallon}$

Answer: _____

What was your driving speed if it took 18 hours to drive this distance (round your answer to the nearest whole number)?

Note: $\dfrac{\text{total miles}}{\text{hours driven}} = \text{miles per hour (speed)}$

Answer: _____

4. According to the map, the shortest route is from _____ to San Francisco.

   a. Houston to Phoenix to Los Angeles
   b. Seattle to Salt Lake City to Los Angeles
   c. Wichita to Salt Lake City
   d. Minneapolis to Seattle
   e. Denver to Salt Lake City to Los Angeles

Reading a Topographical Map

Hikers, backpackers, and forest rangers are some of the many people who are well acquainted with a **topographical map.** A topographical map shows the natural features of an area by means of lines and symbols. The lines on the map are called **contour lines.**

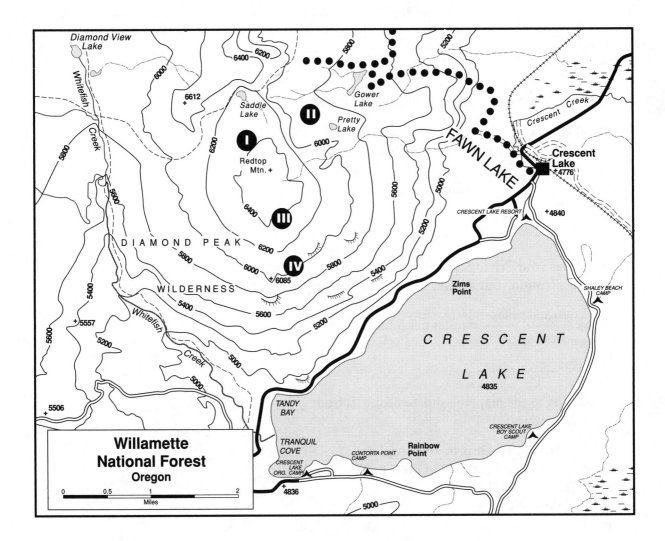

Each contour line represents a specific elevation. The next line over represents a change in elevation, usually 20 feet or more. By looking carefully at the map, you can tell if the elevation increases or decreases from one contour line to the next. You can see that steep terrain is represented by closely spaced contour lines, while flat terrain is represented by widely spaced lines.

Before reading the topographical map, make sure you are familiar with the following diagrams and symbols:

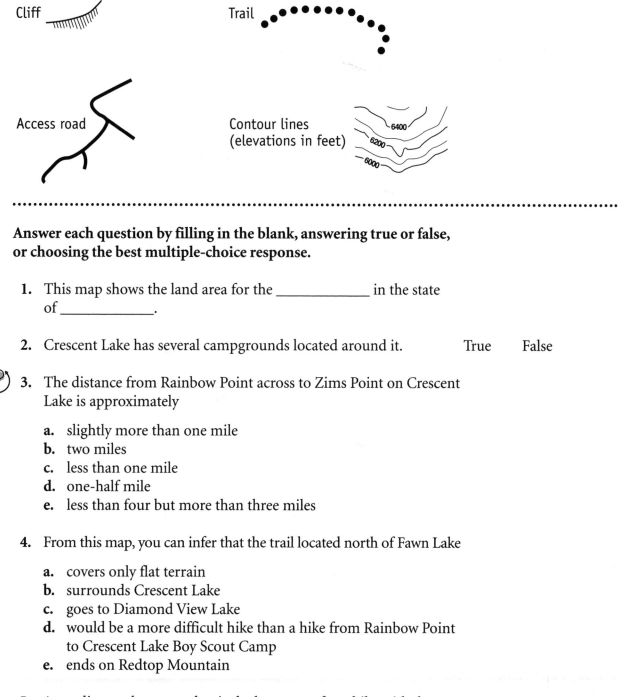

Cliff

Trail

Access road

Contour lines
(elevations in feet)

6400

6200

6000

Answer each question by filling in the blank, answering true or false, or choosing the best multiple-choice response.

1. This map shows the land area for the _____ in the state of _____.

2. Crescent Lake has several campgrounds located around it.   True   False

3. The distance from Rainbow Point across to Zims Point on Crescent Lake is approximately

   a. slightly more than one mile
   b. two miles
   c. less than one mile
   d. one-half mile
   e. less than four but more than three miles

4. From this map, you can infer that the trail located north of Fawn Lake

   a. covers only flat terrain
   b. surrounds Crescent Lake
   c. goes to Diamond View Lake
   d. would be a more difficult hike than a hike from Rainbow Point to Crescent Lake Boy Scout Camp
   e. ends on Redtop Mountain

5. According to the map, what is the best route for a hike with the least amount of exertion?

   a. III to IV
   b. II to I
   c. II to III
   d. I to IV
   e. IV to I

Answer Key

Pages 1–9, Pretest

1. 1998; 1 million
2. True
3. b
4. c
5. Football; Hockey
6. False
7. b
8. d
9. population growth; 2030
10. False
11. c
12. b
13. high school students; 1997
14. False
15. b
16. e
17. average unemployment
18. False
19. d
20. c
21. Chicago, IL; New Orleans, LA
22. True
23. d
24. d
25. Asian Americans
26. True
27. d
28. d
29. Allegheny, Monongahela, Ohio
30. False
31. d
32. c
33. Mexico; Canada
34. False
35. c
36. a

Pages 12–15, Graph Skills Inventory

1. Delta
2. 105 million
3. True
4. False
5. c
6. b
7. widowed; single
8. 8.7
9. True
10. False
11. a
12. c
13. states
14. $31,000
15. False
16. True
17. b
18. b
19. 600
20. True
21. b

Pages 22–23

1. **a.** Average Hourly Earnings in Different Job Categories
   b. Services; Wholesale and Retail Trade; Construction; Manufacturing; Transportation & Public Utilities; Finance, Insurance, & Real Estate
2. $2.00
3. Construction
4. False
5. True
6. False
7. a
8. d
9. d
10. d

Pages 24–25

1. U.S. Department of Labor
2. 1982; 1999
3. percent
4. False
5. True
6. False
7. a
8. a
9. a
10. c

Pages 26–27

1. projected	6. False
2. 10 million	7. a
3. women and men	8. c
4. False	9. b
5. False	10. b

Pages 28–29

1. 2006	6. c
2. 200,000	7. e
3. True	8. c
4. False	9. a
5. True	10. b

Pages 30–31

1. membership	6. c
2. airplane pilots	7. a
3. True	8. c
4. False	9. c
5. True	10. a

Pages 34–35

1. **a.** The Federal Government Dollar: Where It Goes
   b. National Defense; Social Security; Net Interest; Medicare; Medicaid; Non-defense Discretionary; Reserves; Other

2. social security	7. c
3. National Defense	8. e
4. False	9. c
5. False	10. c
6. False	

Pages 36–37

1. motor vehicles	6. True
2. pedestrian	7. d
3. 5	8. c
4. False	9. b
5. False	10. d

Pages 38–39

1. estimated	5. False
2. 322; 292	6. True
3. Housing; Health;	7. a
Defense; Social	8. c
Services	9. d
4. True	10. b

Pages 40–41

1. monthly income	6. d
2. 19%	7. b
3. False	8. a
4. True	9. c
5. False	10. a

Pages 42–43

1. civilian labor force	6. b
2. b	7. a
3. True	8. d
4. False	9. b
5. True	10. e

Pages 46–47

1. **a.** Average Weight for Women 5′ 7″ Tall
   b. average weight in lb
   c. age groups

2. 20; 24	7. c
3. 158	8. c
4. True	9. d
5. False	10. c
6. False	

Pages 48–49

1. average weight	6. True
2. 192 lb	7. e
3. 20; 69	8. a
4. False	9. e
5. True	10. b

Pages 50–51

1. men; women	6. True
2. 20, 24; 70, 79	7. d
3. 160 lb	8. a
4. True	9. d
5. False	10. d

Pages 52–53

1. third party; private consumer	
2. 300 billion	7. b
3. False	8. c
4. True	9. c
5. False	10. c
6. d	

Pages 54–55

1. fatally injured	6. b
2. 40	7. b
3. True	8. a
4. False	9. d
5. d	

Pages 58–59

1. whole milk	6. True
2. dollars; half gallon	7. a
3. 1990; 1997	8. c
4. False	9. c
5. True	10. b

Pages 60–61

1. unleaded premium	6. True
2. cost; gallon	7. c
3. 9	8. d
4. False	9. d
5. False	10. e

Pages 62–63

1. clothing, food, shelter	6. False
2. 1982–1984	7. a
3. clothing	8. e
4. False	9. b
5. True	10. c

Pages 64–65

1. 17	6. b
2. 5%	7. b
3. True	8. c
4. True	9. d
5. False	10. a

Pages 66–67

1. electricity; telephone; gas	
2. electricity; telephone	
3. False	7. b
4. False	8. a
5. True	9. c
6. d	10. e

Pages 68–75, Graph Review

1. education	17. United States; three
2. yearly income	18. 255 million
3. True	19. False
4. False	20. True
5. a	21. d
6. b	22. b
7. c	23. c
8. b	24. d
9. death; U.S.	25. beef; pork
10. 9.6%	26. 1993; $73
11. False	27. False
12. True	28. False
13. a	29. e
14. c	30. a
15. d	31. e
16. e	32. b

Pages 77–79, Schedule and Chart Skills Inventory

1. Central; Mountain	7. 15
2. 3	8. True
3. False	9. False
4. True	10. b
5. d	11. d
6. zone; weight	

Page 87

1. vegetable, milk, meat
2. calories or servings
3. food   7. b
4. False   8. c
5. False   9. b
6. True   10. b

Pages 88–89

1. Johnson Community College
2. Mon., Jan. 26; Sat., Jan. 31
3. 787–1984   7. b
4. True   8. c
5. False   9. c
6. False   10. b

Pages 90–91

1. Suisan-Fairfield; Martinez; Richmond
2. 15; 18
3. sleeping car service
4. True   8. d
5. False   9. b
6. True   10. b
7. d

Page 92–93

1. actual   6. c
2. True   7. c
3. True   8. b
4. False   9. c
5. c

Pages 94–95

1. 25,000; 26,000
2. single; married and filing jointly; married and filing separately; head of household
3. True   7. a
4. False   8. c
5. True   9. e
6. a

Pages 96–101, Schedule and Chart Review

1. mileage   12. False
2. mileage; cities   13. c
3. True   14. d
4. False   15. c
5. a   16. a
6. c   17. car; 7
7. e   18. $10,000; $14,000
8. b   19. True
9. depart   20. True
10. Portland;   21. b
    Seattle;   22. c
    San Francisco;   23. d
    Los Angeles 24. c
11. False

Pages 102–105, Map Skills Inventory

1. mountains   10. c
2. latitude   11. d
3. False   12. 500
4. True   13. annual rainfall; inches
5. c   14. 40; 59
6. downtown   15. True
7. east   16. False
8. False   17. d
9. True   18. d

Page 109

1. northwest   4. southwest
2. southeast   5. south central
3. northeast   6. southeast

Page 111

1. longitude   4. 30° N
2. latitude   5. south
3. equator

Page 113

1. D-3, H-7, D-5   5. Portugal
2. Germany   6. True
3. Baltic Sea   7. False
4. Hamburg   8. False

Page 115

1. **a.** 850   **b.** 450   **2.** 200
   c. 400   **d.** 300   **3.** D-5; 900; southeast
   4. A-1; 850; northwest

Pages 120–121

1. Arctic Circle
2. 500
3. U.S.; Canada
4. 75° W
5. True
6. True
7. False
8. c
9. c
10. e

Pages 122–123

1. Guatemala; Honduras
2. Guatemala; Belize
3. Cuba
4. east
5. True
6. True
7. b
8. b
9. c
10. e
11. c

Page 127

1. San Francisco
2. A-4
3. east
4. False
5. False
6. True
7. True
8. c
9. b
10. b

Page 129

1. A-4
2. Springfield
3. Lake Michigan
4. False
5. True
6. False
7. c
8. a
9. e
10. c

Pages 132–133

1. 2010; Hispanic
2. California; Arizona
3. 500; 999
4. U.S. Census Bureau
5. False
6. True
7. False
8. d
9. a
10. d

Pages 134–135

1. home heating; five
2. electricity
3. A
4. False
5. True
6. b
7. a
8. b
9. b
10. d

Pages 137–141, Map Review

1. 300
2. Mekong
3. False
4. True
5. b
6. d
7. a
8. c
9. 5.3
10. A-7
11. False
12. True
13. d
14. b
15. c
16. b
17. Home Ownership Rates; 1998
18. 60% or less
19. True
20. False
21. a
22. d
23. b
24. b

Pages 142–150, Posttest A

1. U.S. Automobile Corporations; 20 million
2. False
3. d
4. d
5. unemployed
6. True
7. b
8. c
9. 18
10. False
11. c
12. d
13. Northwest
14. True
15. b
16. e
17. 85A; North Central
18. False
19. e
20. d
21. Agriculture
22. False
23. a
24. c
25. Colorado; Arizona
26. True
27. a
28. b
29. G-3
30. True
31. b
32. c
33. aid to dependent children
34. False
35. d
36. a

Pages 152–161, Posttest B

1. d	9. b	17. d	25. a
2. b	10. c	18. b	26. d
3. c	11. a	19. d	27. b
4. c	12. d	20. a	28. d
5. e	13. c	21. c	29. c
6. c	14. b	22. e	30. b
7. b	15. e	23. a	31. e
8. d	16. c	24. b	32. c

Pages 164–165

1. True
2. False
3. e
4. d

5.

INCOME: TAKE-HOME PAY + $2,000		
Expenses	Adjustments (if any)	Corrected Budget
Food	$400	0	$400
Clothing	$170	–$10	$160
Insurance	$80	0	$80
Housing	$660	0	$660
Medical	$140	–$20	$120
Transportation	$145	–$5	$140
Recreation	$200	–$140	$60
Savings	$100	0	$100
Credit Cards	$240	0	$240
Education	$40	0	$40
Total Expenses	$2,175	–$175	$2,000

Pages 166–167

1. $121,300
2. Atlanta
3. d
4. a
5. yes; no; no; no; yes

Pages 168–169

1. $30
2. $60; $120; $180; $180; $280
3. $44; $66; $71.50
4. d

Page 171

1. $3.50; $6.00

2.

Mr. Bill Telephone Pittsburgh, Pennsylvania	Account Number: 123-456		Date Payment Due: January 9, 1999	
CURRENT MONTH'S CHARGES (W = weekday; E = evening; NW = night and weekend)				
In State Calls (direct dial)	Code	Total Time	Amount Charged	Totals
Harrisburg	W	1 minute	$ .44	
Erie	E	2 minutes	$ .47	
York	E	3 minutes	$ .67	
Allentown	W	2 minutes	$ .77	
Harrisburg	W	1 minute	$ .44	
			Total (In State Calls)	$ 2.79
Outside the State (direct dial)				
Los Angeles	NW	3 minutes	$ .69	
Houston	W	1 minute	$ .64	
			Total (Outside the State)	$ 1.33
Operator Assisted (person to person)				
Scranton	W	3 minutes	$ 3.33	
Seattle	W	3 minutes	$ 4.72	
			Total (Operator Assisted)	$ 8.05
Local Telephone Service:	From 11/14 to 12/14			$ 15.15
			TOTAL NOW DUE	$ 27.32

Pages 172–173

1.

	1992		1997
Beef	9.50	Beef	9.00
Milk	5.60	Milk	6.40
Tomatoes	2.40	Tomatoes	3.20
Total	17.50	Total	18.60

2. $1.10
3. Tomatoes and milk
4. Beef: $32.40; Tomatoes: $6.40; Milk: $25.60; Total: $77.40

Pages 174–175

1. Depreciation ___12___ cents
   Repairs/Maintenance ___1___ cents
   Gas/Oil ___$5\frac{1}{2}$___ cents
   Total ___$18\frac{1}{2}$___ cents per mile for the first year

2. $1,850
3. a
4. Depreciation ___4___ cents
   Repairs/Maintenance ___12___ cents
   Gas/Oil ___$5\frac{1}{2}$___ cents
   Total ___$21\frac{1}{2}$___ cents per mile for year 7.

5. a. year 7 b. 0.03 c. i. False ii. True

Pages 176–177

1. $63,000; $52,000; $64,000

2. $43,200; $8,640

3. $500

4. **a.** False   **b.** False   **c.** True

Pages 178–179

1.

From	To	Distance in Miles
Seattle	San Francisco	808
San Francisco	Los Angeles	379
Los Angeles	Houston	1538
	Total Miles:	2725

2.

From	To	Distance in miles	Time in hours (speed = 50 miles per hour)
Seattle	Minneapolis	1608	32
San Francisco	Denver	1256	25
Houston	Raleigh	1160	23
Boston	Miami	1514	30

3. 20.5 mpg; 57 mph

4. **e**

Page 181

1. Willamette National Forest; Oregon

2. True   **4. d**

3. **b**   **5. a**

Using a Calculator

A calculator is a valuable math tool. You'll use it mainly to add, subtract, multiply, and divide quickly and accurately. The calculator pictured at the right is similar to one you've seen or may be using.

To enter a number, press one digit key at a time. On the display, the number 4,863 is entered. Notice the following features:

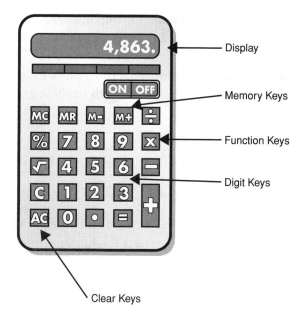

Display

Memory Keys

Function Keys

Digit Keys

Clear Keys

- A decimal point is displayed to the right of the ones digit.

- The calculator does not have a comma key or a dollar sign key.

- Pressing the clear key $\boxed{C}$ erases the display. You should press $\boxed{C}$ each time you begin a new problem or when you've made a mistake.

EXAMPLE 1   Divide 408 by 12.

STEP 1   Press $\boxed{C}$ to clear the display.

STEP 2   Press the digit keys, divide key $\boxed{\div}$, and equals key $\boxed{=}$.

Press Keys	**Display Shows**
$\boxed{C}\ \boxed{4}\ \boxed{0}\ \boxed{8}\ \boxed{\div}\ \boxed{1}\ \boxed{2}\ \boxed{=}$	3 4.

ANSWER: 34

EXAMPLE 2   Multiply $12.75 by 6.

STEP 1   Press $\boxed{C}$ to clear the display.

STEP 2   Press the digit keys, decimal point key, times key $\boxed{\times}$, and equals key $\boxed{=}$.

Press Keys	**Display Shows**
$\boxed{C}\ \boxed{1}\ \boxed{2}\ \boxed{\cdot}\ \boxed{7}\ \boxed{5}\ \boxed{\times}\ \boxed{6}\ \boxed{=}$	7 6.5

ANSWER: $76.50

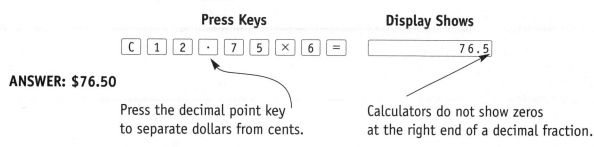

Press the decimal point key to separate dollars from cents.

Calculators do not show zeros at the right end of a decimal fraction.

Using Mental Math

Mental math is the use of thinking skills to help you understand and solve math problems. Estimating in your head is one mental math skill you're already familiar with. Another mental math skill especially valuable in *Number Power 5: Graphs, Charts, Schedules, and Maps* is being able to compare relationships from more than one set of information. Here are some key ideas to help build your mental math skills in these areas.

Graphs

Graphs often require you to compare two or more sets of data.

EXAMPLE 1 How much did sales grow from 1997 to 1999?

In 1997, sales were $6 million and in 1999, sales were $8 million. Subtract 8 − 6 = 2.

ANSWER: Sales grew by $2 million.

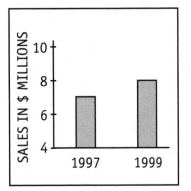

Charts and Schedules

Many charts and schedules challenge you to keep time periods in memory and compute these figures.

EXAMPLE 2 How long does Train 58 take to go from New Orleans to Chicago?

STEP 1 Count the hours within a time frame (A.M. or P.M.). First mentally calculate the time in hours and minutes between 2:15 P.M. to 12:00 midnight (9 hours 45 minutes). Keep this in your mind. Then figure the time from midnight to 9:20 A.M. (9 hours 20 minutes).

STEP 2   Now add these two amounts in your mind and convert
to the next hour if the minutes go above 60.
9 hours 45 minutes + 9 hours 20 minutes =
18 hours 65 minutes = 19 hours 5 minutes

ANSWER: 19 hours 5 minutes

Train #58

READ DOWN	TO
2:15 P.M.	New Orleans, LA
6:06 P.M.	Jackson, MS
1:26 A.M.	Fulton, KY
9:20 A.M.	Chicago, IL

Maps

Maps often require the calculation of miles using a scale.

EXAMPLE 3   How far is it from Los Angeles to San Francisco?

STEP 1   On the map below, every $\frac{1}{4}$ of an inch is equal to
450 miles. Measure the distance from Los Angeles to
San Francisco. You have approximately $\frac{1}{2}$ inch.

STEP 2   Compute this as a multiplication problem. $\frac{1}{2}$ inch is
2 times $\frac{1}{4}$ inch so multiply the distance by 2.
$2 \times 450 = 900$

ANSWER: 900 miles

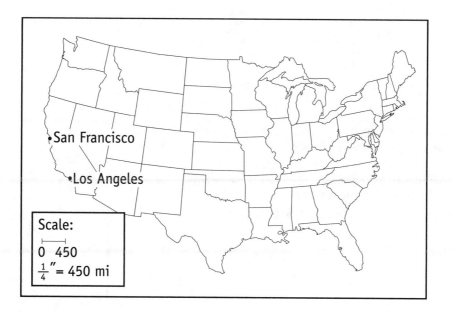

Using Estimation

To estimate is to compute an approximate answer. One way to estimate that works well is to use rounded numbers that are easy to work with. Estimating is an important skill for several reasons. You can estimate to discover "about what" an exact answer should be, to help pick a correct answer from among multiple choice answers, or to quickly check an answer obtained when using a calculator.

In *Number Power 5: Graphs, Charts, Schedules, and Maps,* there are some questions for which you are asked to estimate an answer. For example, you may need to use estimation when computing miles on a map or computing a number on a graph. On these questions, you do not need to calculate an exact answer. Here are some general guidelines for estimating answers to problems you'll find in this book.

Graphs

Often with some line and bar graphs, you are asked to determine a pattern or trend.

EXAMPLE 1 According to this graph, estimate how much sales will grow each year.

Sales in 1997 are almost halfway between 6 and 8 million, so you can estimate them to be about $7 million. If sales are $7 million in 1998 and $8 million in 1999, you can estimate sales to grow by $1 million each year.

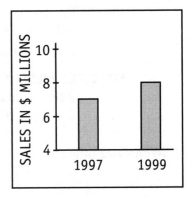

EXAMPLE 2 If profits continue to grow at the same rate each year, what will profits be in 2000?

The profit in toys continues to grow by about $2 million each year. You can estimate that profits in the year 2000 will be about $10 million.

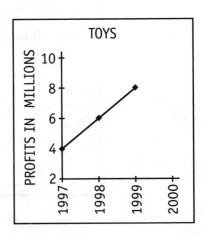

Charts

EXAMPLE 3 If you eat 1,200 calories per day, estimate how many sandwiches you can eat to stay within this diet.

For Big Macs you can eat eight (1,200 ÷ 150). For Wendy's cheeseburger, you can estimate 4 sandwiches (1,200 ÷ 300).

	ITEM	Calories	Fat(g)
McDonald's	Big Mac	150	35
Wendy's	Jr. Cheeseburger	320	13
Burger King	Whopper w/cheese	730	46

Maps

EXAMPLE 4 Use the mileage scale to estimate the distance from Los Angeles to New York City.

On the map below, every inch is equal to approximately 600 miles. The distance from Los Angeles to New York City is about 4 inches or 2,400 miles (600 × 4).

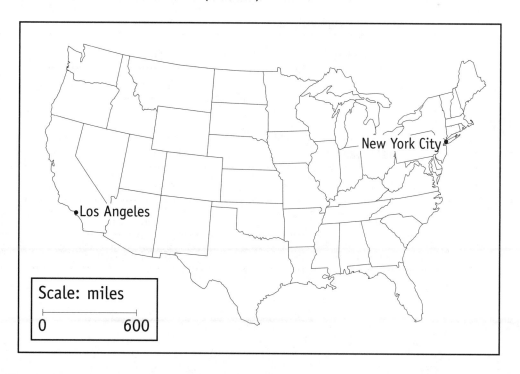

Measurement Conversions

LENGTH

Customary	Customary	Metric
1 inch	$\frac{1}{12}$ foot or $\frac{1}{36}$ yard	2.54 centimeters
1 foot	12 inches or $\frac{1}{3}$ yard	0.3 meter
1 yard	36 inches or 3 feet	0.91 meter
1 mile	5,280 feet or 1,760 yards	1.61 kilometers

LIQUID

Customary	Customary	Metric
1 ounce	$\frac{1}{16}$ pint	29.57 milliliters
1 cup	8 ounces or $\frac{1}{2}$ pint	0.24 liter
1 pint	16 ounces or 2 cups or $\frac{1}{2}$ quart	0.47 liter
1 quart	2 pints or 4 cups or $\frac{1}{4}$ gallon	0.95 liter
1 gallon	8 pints or 4 quarts	3.79 liters

WEIGHT

Customary	Customary	Metric
1 ounce	$\frac{1}{16}$ pound	28.35 grams
1 pound	16 ounces	453.6 grams
1 ton	2,000 pounds	907.18 kilograms

TEMPERATURE

°C = (°F − 32) ÷ 1.8

°F = (°C × 1.8) + 32

Glossary

A

abbreviation   One or more letters that are used to represent longer words. Directions on a map are usually abbreviated: N stands for north, S stands for south, E stands for east, and W stands for west.

annual   Referring to anything that occurs each year

average   A typical or middle value of a set of values. The arithmetic average (mean) is found by adding the set and then dividing the sum by the number of values in the set.

Stacey received three math scores: 83, 78, and 91. Stacey's average score is 84.
83 + 78 + 91 = 252
252 ÷ 3 = 84

axes   The sides of a graph along which data values or labels are written

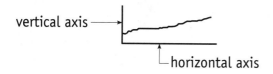

vertical axis ⟶

horizontal axis

B

bar graph   A graph that uses bars to display information. Data bars can be drawn vertically or horizontally.

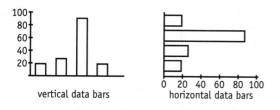

vertical data bars   horizontal data bars

budget   A financial plan that details expenses and sources of earning or revenue

C

chart   A graph or drawing that contains data or other information

circle graph

circle graph   A graph that uses a divided circle to show data. Each part of the circle is called a *segment* or a *section*. The segments add up to a whole or to 100%. A circle graph is also called a *pie graph* or *pie chart*.

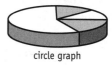

circle graph

column   In a table, a vertical listing of numbers or words that is read from top to bottom

column of numbers	column of words
124	Newport
230	Jamestown
195	Cleveland
324	Detroit

continent   One of the 7 major land masses of the earth. The continents are Antarctica, Africa, Asia, Australia, Europe, North America, and South America.

contour lines   Lines on a topographical map that represent points of equal elevation

D

data   A group of numbers or words that are related in some way

Number data: $2.50, $3.75, $6.40
Word data: beef, chicken, fish, pork

diagram   Any chart or graph that contains numbers, words, or other information

direct distance   The straight-line distance between two points as might be flown by an airplane

distance scale   A map scale that relates ruler distance on a map to actual mileage

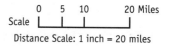

Distance Scale: 1 inch = 20 miles

double bar graph   A bar graph containing two sets of bars to display and compare two sets of related data

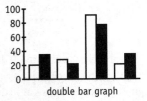

double bar graph

double line graph   A line graph containing two lines to display and compare two sets of related data

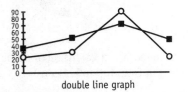

double line graph

directional map   A map that is used to show the location of cities, highways, and points of interest. Directional maps are often called *road maps*.

E

east   A map direction (**E**) that is opposite of west. When north is placed at the top of the direction symbol, east is placed at the right side.

Eastern Hemisphere   The part of the earth that includes Asia, Africa, Australia, Europe, their islands, and all surrounding water

elevation key   A symbol or number, often on a contour line, that tells land elevation

equator   An imaginary line that circles the earth as a circumference, lying halfway between the North and South Poles. The equator is the line of 0° latitude.

extrapolate   To make a reasonable guess of a data value that lies outside a given set of values

Four given data values: 5, 10, 16, 20
Extrapolated fifth value: 25

G

geographical map   A map that shows the natural features of the earth. Natural features include lands, rivers, lakes, oceans, and other features that are not man-made.

globe   A map in the shape of a sphere, showing the whole earth

Globe

graph   A pictorial display of information. A graph usually includes a graph title, data, horizontal and vertical axis labels, and the graph's source of information.

Greenwich, England   A town in England through which the 0° longitude line passes. The longitude line numbered 180° is exactly halfway around the earth from Greenwich.

H

hemisphere   On a map or globe, the word *hemisphere* refers to approximately half of the earth.

horizontal   On a graph, the direction left to right or right to left. On a map or globe, the direction west to east or east to west

horizontal axis   On a graph, the axis running from left to right

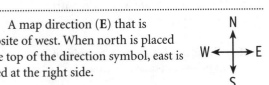

horizontal axis

I

index guide   A set of map coordinates that is used to locate a point on a map. The coordinate of each point is usually given as a letter followed by a number—the letter giving the north/south location, the number giving the east/west location.

inference   A guess at information that is not directly given

informational map   A map that gives specific information about a particular region, or gives information comparing different regions

interpolate   To estimate a data value that lies between two given values

K

key   On a pictograph, a explanation of the meaning and value of a data symbol

L

latitude lines   Lines drawn horizontally (west to east) across a globe. The latitude line number tells how far a point on the earth is north or south of the equator.

line graph   A graph that displays data as points along a graphed line

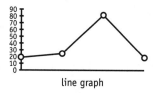

line graph

longitude lines   Lines drawn vertically (north and south) on a globe. A longitude line number tells the east/west location of a point on the earth.

M

map   A visual display that represents a city or a region of the earth

Mercator projection   A flat map of the whole earth on which longitude lines are drawn as parallel lines. Exact locations are shown but sizes become increasingly distorted with distance from the equator.

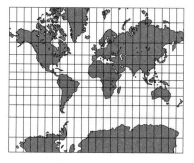

N

north   A map direction (N) opposite of south. North is usually placed at the top of the direction symbol.

Northern Hemisphere   The half of the earth that lies north of the equator

North Pole   The northern most point on the earth. On a globe, the North Pole is at the top of the globe.

northeast   Any direction that is both north and east of a point

northwest   Any direction that is both north and west of a point

O

ocean   The large body of salt water that covers most of the surface of the earth. The ocean is often geographically divided into five major oceans: Antarctic, Arctic, Atlantic, Indian, and Pacific. The Pacific Ocean borders the western shore of the United States; the Atlantic Ocean borders the eastern shore.

P

percent   Part of 100. For example, 5 percent means 5 parts out of 100, 5 hundredths of the whole.

pictograph   A graph that uses small pictures or symbols to represent data. Data lines may be displayed either horizontally or vertically.

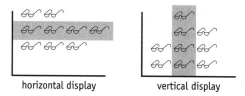

horizontal display   vertical display

pie graph   See *circle graph*. (Pie graph is another name for circle graph.)

prediction   Regarding a graph, a guess about the value of an unknown data point

R

road distance   The actual road mileage distance between two points

row   In a table, a horizontal list of numbers or words that is read from left to right

row of numbers: 26, 32, 65, 124
row of words: carrots, celery, lettuce, sprouts

S

schedule   A list of facts and relations primarily dealing with times of events

```
        CITY CENTER BUS STATION
        WEEKEND TIME SCHEDULE

  Saturday   8:00 A.M.   10:30 A.M.   1:00 P.M.
  Sunday     9:00 A.M.   11:30 A.M.   2:00 P.M.
```

sea level   The average level of the ocean from which altitude measurements are compared. Land that is the same level as average level of the ocean is said to have a *zero elevation.*

segment   On a circle graph, any pie-shaped division

←— segment

source   On a schedule, chart, graph, or map, a telling of where information was obtained

south   A map direction (S) that is opposite of north. South is usually placed at the bottom of the direction symbol

southeast   Any direction that is both south and east of a point

Southern Hemisphere   The half of the earth that lies south of the equator

South Pole   The southern most point on the earth. On a globe, the South Pole is at the bottom of the globe.

southwest   Any direction that is both south and west of a point

symbol   An icon or other drawing that represents a number, word, or object. For example, on a topographical map these symbols are used:

cliff   trail

T

table   A display of data (words and numbers) organized in rows and columns

Name of School	Women Teachers	Men Teachers
Jefferson	14	6
Lincoln	12	9
Washington	17	4

row ——→ Lincoln   column

title   On a graph, a short description of the main information presented

topographical map   A map that shows natural land and water features, showing elevation by means of labeled contour lines

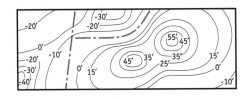

V

vertical axis   On a map, the axis running up and down

vertical axis

W

west   A map direction (W) that is opposite of east. If north is placed at the top of the direction symbol, west is placed at the left side.

N
W — E
S

Western Hemisphere   The part of the earth that includes North and South America, their islands, and the surrounding water

Index